Praise for *This Is What*

"I learned a lot from this book—and w
—Dr. Daniel J. Levit<mark>in, *New Yor*</mark>
This Is Your Brain on Music

"If you've ever wondered why you love a song and what that says about you, this book will help you understand why."
—Touré, author of *I Would Die 4 U:*
Why Prince Became an Icon

"Why do we like the music we like? With a provocative blend of studio stories and fascinating neuroscience, celebrated producer and engineer Susan Rogers sets out to answer this eternal mystery—and, along the way, just might turn you into a better listener."
—Alan Light, music journalist and author of *Let's Go Crazy:*
Prince and the Making of Purple Rain

"A groundbreaking study of great intervention. . . . [T]his is the book that scholars and fans of popular music across all disciplines have impatiently waited for."
—Stan Hawkins, professor of musicology, University of Oslo

"Susan Rogers is one of the greatest listeners and feelers of music I've met, and that's what makes her such a great record producer. . . . [S]he has opened my mind to how we hear music and has once again inspired me as both listener and musician."
—Steven Page, Barenaked Ladies

"Susan Rogers's words dance on the page with their sheer enthusiasm and eloquence. The way she illuminates what makes music so effective—from breaking down a Kanye West instrumental to the vocal skill of Frank Sinatra—will have you reconsidering songcraft and the way you process it."

—Kate Hutchinson, journalist and broadcaster

"I understood *why* I love the records I do and now I've got a map for my next treasure hunt."

—Mobeen Azhar, award-winning journalist and filmmaker

"The mysterious gravitational tractor beam of musical obsession that has directed my entire life has now been explained so eloquently by Susan Rogers. She has guided me on a righteous riff to the engine room of my own unique musical journey."

—Craig Northey, musician and film and television composer

"[Susan Roger's] passionate love of music, her vast experience creating history in the recording studio, and her deep knowledge about the subject helped me understand WHY I love the music in my life."

—Duane Tudahl, author and television producer

THIS IS
WHAT IT
SOUNDS
LIKE

A Legendary Producer
Turned Neuroscientist
on Finding Yourself
Through Music

Susan Rogers
and Ogi Ogas

W. W. NORTON & COMPANY
Celebrating a Century of Independent Publishing

For information about permission to reproduce selections from this
book, write to Permissions, W. W. Norton & Company, Inc.,
500 Fifth Avenue, New York, NY 10110

For information about special discounts for bulk purchases,
please contact W. W. Norton Special Sales at
specialsales@wwnorton.com or 800-233-4830

Manufacturing by Lakeside Book Company
Book design by Lovedog Studio
Production manager: Anna Oler

The Library of Congress has catalogued the hardcover edition of this
book as follows:

Library of Congress Control Number: 2022030815

ISBN 978-1-324-06596-8 pbk.

W. W. Norton & Company, Inc.
500 Fifth Avenue, New York, N.Y. 10110
www.wwnorton.com

W. W. Norton & Company Ltd.
15 Carlisle Street, London W1D 3BS

1 2 3 4 5 6 7 8 9 0

For Art, whom music saved

That sense of absolute freedom, that sense
of no direction but the greatest direction in
the world, of being able to feel,
I'm a part of this somehow.

—*Sam Phillips, producer,*
on pursuing a life in record making

CONTENTS

LINER NOTES

Listening to This Book

This book invites you to listen to many specific records. For the best experience, we recommend that you search for and listen to these records on a popular music streaming service such as Spotify, Tidal, Apple Music, Pandora, iHeartRadio, or Amazon Music. You can find a list of *This Is What It Sounds Like* records on the website we've created, ThisIsWhatItSoundsLike.com.

THIS IS
WHAT IT
SOUNDS
LIKE

OVERTURE

I CAN IDENTIFY THE EXACT MOMENT WHEN MY JOURNEY to becoming a professional music listener began. It was at a Led Zeppelin concert at the Forum arena in Los Angeles when I was twenty years old. Hundreds of concerts later, I'd still rank it as one of the best I've ever seen. Robert Plant was at the height of his fame as a rock god, mesmerizing the crowd with his vocals, while guitarist Jimmy Page, clad in a black silk suit embroidered with orange-and-crimson dragons, struck incendiary power chords. But as the concert approached the halfway point—the band still hadn't performed classics like "Kashmir" and "Stairway to Heaven"—I realized that it was time for me to leave.

It broke my heart to go. Music was the truest source of passion and meaning in my life and the concert had lifted me to a state of pure rapture. But if I wasn't home by ten-thirty p.m., there would be hell to pay. Not from my parents. My mother had died when I was fourteen years old, and I no longer lived with my father. I had dropped out of high school at age seventeen and married an older boyfriend, thinking that matrimony would be a quick ticket to security and independence. Instead, my marriage had become a trap of desperation and loneliness. My husband resented my attraction to music, and if I didn't return home by his curfew, I would be met at the door with jealous wrath—or worse. So as Page launched into the acoustic arpeggios that open "Bron-Y-Aur Stomp," I apologized to my bewildered friends and solemnly made my way to the exit.

I felt so powerless and low. All my life I had enjoyed an intense, irresistible, and what seemed like a *necessary* relationship with music—when I listened to music, every note felt important and every lyric felt true—yet I'd been intimidated into abandoning one of the most exhilarating musical experiences of my life to return to a place of isolation. Unexpectedly, defiance overtook me. I stopped in my tracks and, in dramatic Scarlett O'Hara style, leaned back, raised my eyes toward the rafters, and vowed, *One day I will return to the Forum and mix live sound for an amazing band!*

It was an utterly implausible vow. For starters, I wasn't entirely sure what a sound mixer actually did, let alone how to become one. I couldn't play an instrument. I didn't sing. I didn't know any musicians or anyone in the music industry. I stitched heart valves on a biomedical assembly line for a living. Yet this improbable musical fantasy had been germinating in the back of my mind ever since I'd seen a photograph on the back of a Sonny & Cher album when I was very young. It showed a man sitting in front of an elaborate console of knobs, buttons, and sliders. The label below him read, "Sound engineer." When I saw this photo, I felt more than thought, *He's making records but not playing an instrument—maybe I could do that!*

Not long after that Led Zeppelin concert, I decided to act on my vow. I divorced my husband and moved to Hollywood with less than a hundred dollars in my bank account. I persuaded a professional audio company to take a chance and hire me as an audio technician trainee. They taught me to install and repair the sophisticated electronic equipment in recording studios, the magical workshops where musical wizards made albums. Working as an audio tech wasn't nearly as glamorous as making music—or mixing it, for that matter—but in a few years' time it provided me with an intimate view of record making by talented artists such as Crosby, Stills & Nash, Jackson Browne, and Bonnie Raitt.

I admired all the artists whose music I was fortunate enough to serve, but my favorite kind of music was soul: performers like James Brown, Marvin Gaye, Al Green, and Sly Stone. And there was no question which artist was my favorite. The newcomer mowing down all the conventions of rock, pop, and soul: *Prince*. But Prince recorded his music elsewhere, and I knew that he wasn't ever going to walk through the door of the Los Angeles studio where I worked.

Then one life-changing day in the early summer of 1983, I got a call from a former boyfriend who worked as chief tech for Westlake Audio, Michael Jackson's chosen studio. In his thick Boston accent John said, "Yah dream job is waitin' fah ya—Prince is lookin' fah a technician!" Instantly I knew that the job would be mine. I had been a Prince fan from the moment I'd heard his first single, "Soft and Wet," coming out of a boom box held on the lap of a young Black teen on the back of an east-bound bus crawling along Sunset Boulevard in Hollywood. I had seen Prince a couple of times on tour and had all his albums. I also knew through the professional grapevine that he liked working with women.

His genre-bending sound and avant-garde profile made him a rare bird, an eclectic outsider to the folk-rock musicians surrounding me in Southern California. But the fact that I was one of a tiny handful of female audio technicians in the industry made me a rare bird, too. None of the more experienced technicians in Los Angeles were interested in taking the gig, because it required relocating to the Midwest, thousands of miles from the epicenter of the entertainment industry. I, on the other hand, was willing to leave everyone and everything behind, move to Minneapolis, and become the personal tech for the artist who meant the most to me.

Prince had just finished touring for his groundbreaking double album *1999*. He was at an early point in his next record, and the first thing he needed me to do was install a new recording console in his home. That task took me about a week to finish, after which we began to have short exchanges. (Prince was famously taciturn, especially while working.) What little conversation we did have was about equipment and practical matters, but one day as I drove up to the gate outside his house I had Sly & the Family Stone's "Thank

You (Falettinme Be Mice Elf Agin)" cranked up on my car stereo. I lowered the car window and pressed the button for the intercom. Prince answered . . . by singing along.

That might have been the first time he realized that he and I shared the same musical taste—that we "lived on the same street," as he liked to put it. Perhaps that's why after I finished installing the equipment, he unexpectedly invited me to sit in the recording engineer's chair.

There's a world of difference between an audio tech and a recording engineer. To use an analogy from the movie business: the recording engineer is like the cinematographer, while the audio tech is like the crewperson who repairs the camera. But Prince didn't realize— or, more likely, didn't care—that I had no experience in the art of manipulating sound. He trusted my technical knowledge and, somewhat audaciously, was willing to trust my ability to listen. For someone so determined to make her way in the music industry, it was an astonishing opportunity. I was learning to make records from one of the most original musical minds of his generation, hearing what he heard, listening through his ears. My first professional experience in the engineer's chair helped me realize that all the happy hours I'd spent listening to records had supplied me with a rich mental framework for making musical decisions. Once an engineer masters the technical tools of the trade, the next step is to use those tools to make something she likes. My lifelong love—listening to records— helped me push sound around and arrive at something my new boss approved of.

The album we worked on in that basement studio was *Purple Rain*. It became one of the most successful and influential albums of all time, selling more than twenty-five million copies and earning Prince two GRAMMYs and an Academy Award. It stayed at No. 1 on the *Billboard* Top 200 chart for twenty-four weeks, tied for the sixth-longest run in history.

I accompanied Prince on the *Purple Rain* tour as a sound engineer, and one of our tour venues was the Los Angeles Forum. Though I wasn't mixing sound at the "front of house" console, I had what I considered an even better gig: I was recording the show in a mobile audio truck parked behind the stage. This show would be captured for posterity, and I was in charge of it.

In most respects, this was a routine tour date. The light and sound systems were rigged and tested. The risers and stage props were placed. The instruments and microphones were set up, routed, and line-checked. All this was accomplished by the efficient hands of a well-trained road crew. But this particular tour date was anything but routine for me.

I went through the usual motions of my tasks: mic adjustment, gain staging, and meticulous sound sculpting during the band's soundcheck. But all the while I could *feel* those rafters high above my head. I remembered exactly where I'd stood in the stands eight years before, when I'd made my impossible vow. Now it was coming true: I was a professional recording engineer working at the Forum for my favorite artist in the world.

A few hours before the show, I took the cassette from our sound-check into Prince's dressing room to touch base on some last-minute details. He was sitting alone at a makeup table. Although I rarely divulged anything personal to him, this moment meant too much to me to keep it to myself.

"Prince, I want to tell you something . . ."

He turned to me with an inquisitive look. I told him about my Led Zeppelin concert vow and capped off my narrative with what I hoped were plain words expressing a simple truth: " . . . and I just wanted to thank you for making my dream come true."

Prince was widely known as someone who avoided small talk and unnecessary personal conversations, preferring to share his thoughts in his lyrics. Whether in public or working with collaborators, he maintained a poker face to protect his artistry from the relentless distractions of stardom. I seldom saw the true, pre-celebrity Prince under this veil, but tonight was one of those rare moments. A broad smile lit up his face, expressing genuine joy at making my fantasy come true—and in his eyes I could recognize that he, too, was living out a dream. Nothing else needed to be said. This was one of our most resonant moments during four extraordinarily productive years together.

In time, I returned to Los Angeles and began working as an independent recording engineer. Soon, record labels and artists began to recognize that I had a good ear along with technical skills. My career took another step forward as I transitioned into the role of record producer. If a recording engineer is like a movie's cinematographer, then the record producer is like the director, guiding performances, critiquing arrangements, and shaping the finished product to meet its artistic goals.

By the mid-1990s, I had become one of the very few successful female record producers in the profoundly male-dominated industry. (During most of my production career in the 1980s and '90s, you could count on the fingers of one hand the number of women who worked exclusively in a production role.) I engineered, mixed, and/or produced music with a variety of artists, including David Byrne, Tevin Campbell, Rusted Root, Robben Ford, and Geggy Tah. I worked on gold and platinum albums, including co-producing the *Billboard* No. 1 hit "One Week" with Barenaked Ladies. I had reached a pinnacle of professional music listening: turning songs into *records*—the objects of listeners' devotion and affection. Yet for much of my career, even though I was working closely with some of the most talented musicians in the business, in my private moments I often found myself asking an unsettling question.

Did *listening* to music actually matter?

Though I was well paid and consistently earned the respect of my colleagues, I felt there was an invisible line in the studio that I could not cross. In my mind, I remained "merely" a listener. I could often be swayed by what the musicians suggested, even when I held a contrary view, because I believed that my own musical opinions didn't carry the same weight as a composer's or performer's. I embraced and cherished the role of record maker, yet during the more nuanced conversations of how music worked, I sometimes felt like a bystander.

My beliefs about the importance of listening did not change overnight, but I can trace the start of the shift to a memorable encounter with one of the most celebrated musicians of all time: jazz legend Miles Davis. Though it took several years to fully bloom, a germ of wisdom he planted would eventually transform my perspective on what music listening was ultimately all about—and lead me to write this book.

One day Miles was visiting Prince's Minnesota home for dinner and to listen to a few of our works in progress. After they finished eating, they came downstairs to the studio, where I was waiting. Miles was standing with his back directly in front of me as he addressed Prince's father, John Nelson, a jazz pianist. The two veteran musicians were in the middle of an odd dinner conversation about pants.

"I really like those striped pants you wear," said Prince's dad.

"I don't wear striped pants," Miles replied.

"Yes, you do—I've seen you wear them."

"Where'd you see me in striped pants?"

"On TV. At the GRAMMYs."

"I don't have any striped pants!"

"Yes, you do! They're black and white!"

"I'm telling you, I don't have any striped pants!" insisted Miles.

All of a sudden, he spun around and thrust his famously intense globular eyes just inches from my face and exclaimed, "Yes, I do!" as if he had been talking to me all along. "They're made out of eel, like in Vietnam!"

Holding my ground, I blurted, *"EEL? Like in Vietnam?"*

Standing rock still, Miles Davis kept his face uncomfortably close and began firing off a rapid barrage of questions:

"Who're you?"

"Susan."

"Where you from?"

"Anaheim."

"What do you do?"

"I'm an engineer."

"How long you been here?"

"A few years."

As the barrage continued, it gradually dawned on me that this odd exchange was what jazz musicians do: Miles was "playing" verbal riffs for me to respond to, going back and forth without pause, as if improvising with a fellow instrumentalist. I took a modest amount of pride in returning his serves and keeping up with the pace. At the climax of our duet, Miles fired off the proclamation that would inspire me to reconsider the value of my contribution to music.

"You a musician?"

"I am not."

"That's okay," he declared, his big, round eyes still locked onto mine. *Some of the best musicians I know aren't musicians.*

And with that he turned away, impromptu jam session over.

I didn't fully comprehend Miles's enigmatic claim that day, but I felt that he had said something that was both true and crucial to understanding the nature of music. His words hovered around my consciousness as I worked with musicians, producers, and engineers, exploring music from the minuscule details to the biggest of big-picture concerns. As I gained confidence in the studio and watched a tremendous variety of musical minds at work, I slowly came to realize that I was developing a perspective that complemented the formal music theories taught at conservatories like Juilliard and Berklee.

Listening to music—attending to what works and doesn't work in a song, feeling its rhythms and melodies as if they were as much a part of your body as your fingers and hips—is an indispensable component of *what music is*. Practically speaking, without a listener, music does not exist. By perceiving, feeling, and reacting to the many dimensions of a song, a listener closes the creative circle and completes the musical experience. Implicit in Miles Davis's declaration was the conviction that *when it comes to the creation of a musical experience*, the act of listening can be every bit as vital as the act of performance.

But note that crucial qualification: *can be*. It doesn't happen automatically. Listening is *not* the same as hearing. Listening is an active process, not a passive one, and becoming a competent musical listener requires curiosity, effort, and love. But it requires something else, too—something it took me years to fully comprehend. *It requires understanding and embracing your unique identity as a listener.* That's why I wrote this book: to help you better understand "the street you live on" so that you can get more out of your relationship with music, even if (like me!) you can't tell an A-sharp from a B-flat.

We each seek out different sorts of experiences and emotional rewards from our musical encounters. Some listeners favor songs that evoke sweet nostalgia, while others crave a groove that matches their inner rhythm. Some listeners prefer to let their imaginations wander freely when they enjoy their favorite records, while others visualize specific scenes evoked by a song's lyrics. Some listeners covet innovative sound design, while for others it's all about that bass. Science simply cannot predict how a person will respond to a piece of music from its objective features alone. Music's *features* do not predict love—*music listening* does. Two people can listen to the exact same song and report dramatically different accounts of "This is what it sounds like . . . *to me.*"

This book is premised on the notion that these divergent reactions are not due primarily to your level of musical training, your circle of teenage friends, or even the year you were born. The music that delivers the maximum gratification to you is determined by seven influential dimensions of musical listening: authenticity, realism, novelty, melody, lyrics, rhythm, and timbre. Collectively, your natural response to each of these seven dimensions forms a personalized "listener profile" unique to you. Your listener profile determines your thoughts, feelings, and physical responses when you listen to music.

The dimensions of your listener profile serve as distinctive routes through which your body and brain can fall in love with a piece of music. Each dimension contains a personal, neural "sweet spot" where music can provide you with your deepest experience of musical joy. Your melodic sweet spot may be tapped by songs in a minor key with a sad refrain, while your friend may relish tunes with brisk, uplifting melodies. Your sweet spot for rhythm might be the hopping beats of ska or reggae, while mine might be the rolling bass of R&B. To help you recognize and appreciate your own sweet spots, I will invite you to listen to records across all manner of genres—including artists you might imagine don't make "your kind of music."

By gaining a better understanding of your musical identity as a listener, you will deepen your connection to music, feel empowered to live a more musical life, hear the music you've always loved with fresh ears, and—I hope—learn something new and surprising about yourself.

So then. As Prince liked to say when he was done with the preliminaries and ready to kick things off:

Band, on stage!

Records Versus Songs

I will be using the term "record" to refer to a specific physical recording of a piece of music, which could be on a vinyl disc, CD, YouTube video, streamed audio file, or Playskool Tape Recorder. In contrast, I'll use the term "song" to refer to the words and melody of a particular piece of music, regardless of who is performing or recording it.

CHAPTER 1

AUTHENTICITY

This Is What Expression Sounds Like

The wrong note played with gusto always sounds
better than the right note played timidly.

—*Tommy Jordan,*
lead singer of Geggy Tah

1

SOME OF THE HAPPIEST TIMES IN MY LIFE HAVE BEEN spent in record pulls. A record pull is when a group of friends or colleagues get together and play music for one another. There are two rules. First, you must choose records that are personally meaningful to you. It could be a great performance or eye-opening lyric, a record you associate with a significant time in your life, or even the record you want to dance to at your wedding. The second rule is a bit more of a guideline. Ideally the record should be something unfamiliar or uncelebrated—something your companions probably don't know well—though this rule can be broken if you want your pullmates to hear a popular song in a new way.

One of the happiest outcomes of a record pull is the opportunity to discover something new about the way your companions think and feel. Gaining insight into a friend's musical tastes can be an intimate experience that reveals how they see themselves in relation to the world, the value of aesthetic experiences in their lives, or who they want to be when they grow up (or who they wanted to be). It's not merely a pullmate's choice of songs that is revealing but their explanation of *why* a record matters so much to them. Good record pulls feature as much storytelling as music.

I once shared a record pull with music-cognition scientist Daniel Levitin, author of *This Is Your Brain on Music*, and he chose to play a once-popular record by Michael Nesmith of the Monkees, "Joanne." Though I had heard the song many times before, after Levitin offered a surprisingly heartfelt account of the record's influence on his own early career as a songwriter, I came to recognize a tenderness I'd never noticed before in the earnest and folksy pop tune—and in Dan's own musicality.

A record pull is also a terrific way to learn something new about yourself. By singling out records that express your deepest identity, by taking the risk of others hearing your private musical passions, and by describing why you fell in love with a record in the first place, you can get in touch with the nuances of your musical self. The reciprocity of a record pull also exposes you to new music that you may not have encountered, draws your attention to details that you might not have appreciated before, and lets you compare your own responses to those of your friends. The best record pulls are not merely social gatherings but adventures in self-discovery.

This book is designed as a kind of record pull, but one with a specific aim in mind. Each chapter contains records intended to help focus your attention on how *you* connect with music. A few of these selections have personal meaning to me or my coauthor, the neuroscientist Ogi Ogas. As you will soon discover, the two of us are very different kinds of listeners—yet our taste in music is beside the point. My hope is that by contrasting our divergent responses to music, we'll help you get better acquainted with your *own* musical identity—especially those hidden aspects of your musical appetites that you may have never recognized before.

I'd like to kick off our *This Is What It Sounds Like* record pull with a track by one of the most controversial bands in American music, celebrated by music industry insiders as a unique talent . . . though they've also been dismissed as "hauntingly bad." I'm sharing their record with you because their music represents one of the purest examples you will ever hear of *authenticity*. But before we can appreciate the lessons offered up by these girls' odd music, we must first hear their odd story . . .

2

The tale opens with an obsessive father by the name of Austin Wiggin Jr., a mill worker in the rural town of Fremont, New Hampshire. Before having children, Austin lived his entire life without any meaningful interest in music, other than occasionally twanging a jaw harp. The origins of Austin's peculiar role in the history of music do not lie within any personal passion for the art of sound, but in a soothsayer's prophecy.

Austin's mother was a farmland oracle. When her son was a young man, she prognosticated that one day he would marry a woman with strawberry-blond hair—and lo and behold, so he did. She also had a premonition that after she died, two sons would be born unto Austin—and sure enough, after she passed on, Austin and his wife were blessed with two boys. His mother's final augury was that his three daughters would one day form a mighty band whose music would be celebrated throughout the land. But for this last and greatest prophecy to come to pass, Austin believed that he would need to guide the hand of providence. So, in the late 1960s, Austin Wiggin set out to transform his daughters from a trio of provincial and somewhat frumpy teens into a female version of the Beach Boys.

The three Wiggin girls were Dot, Betty, and Helen, their names as ordinary and down-to-earth as their hometown. Austin assigned Dot to vocals and lead guitar. Betty was directed to vocals and rhythm guitar. Helen was appointed to drums. Austin knew that attaining their foredestined renown would require total devotion to the cause. He removed the girls from school and took complete charge of their education. He forbade them from seeing friends or dating boys. He didn't even allow them to listen to popular music, out of fear that it might taint their natural gifts. Instead, he compelled them to practice all day, every day, guided by his own judgment of what constituted musical excellence.

Cut off from the world in their pastoral New England home, quietly obeying the commands of their father, the Wiggin sisters' lives didn't resemble the conventional backstory of rock 'n' roll idols. They were more like a prim troika of Emily Dickinsons than a marauding threesome of Joan Jetts.

"We weren't allowed to go out or go to dances or anything. We just stayed at home. He didn't want us to get too involved with the outside," Dot Wiggin would later say in a BBC interview. "We practiced during the day while he worked, then we would practice when he came home from work, sometimes we'd practice before supper. We practiced until it was the way he liked it. If he didn't like it, we'd do one song over and over and over."

After they'd spent years working on their craft ("honing" is not the appropriate verb) and playing regular Saturday concerts at the Fremont Town Hall for the children of farm hands and mill workers, Austin thought they were finally ready for their studio debut. He named his daughters' band the Shaggs, perhaps after the shag haircut that was popular at the time (and which the girls sported for a while) or perhaps after the shaggy dogs beloved by the girls. He booked time at a major recording studio near Boston so that they could record the twelve songs they had written and assemble them into their first album. The recording engineer at the studio, Russ Hamm, became the first industry professional to hear the Shaggs perform. And the moment the girls began to play, he knew exactly what it sounded like.

Incompetence. Embarrassing, unsalvageable, breathtaking incompetence.

Staff musician Bob Hearn joined his colleagues in the control room and described their reaction: "We shut the control room doors and rolled on the floor laughing. Just rolled! It was horrible. They did not know what they were doing but they thought it was okay. They were just in another world."

You will probably understand Hearn's reaction when you listen to the Shaggs track I'm about to discuss. It starts out with uncertain chords played on guitars that sound out of tune even to the most untrained ears. Each strum is preceded by an unexpected rhythmic pause that seems to suggest some reluctance on the part of the poor instruments. Next, the ragged beat of a snare clambers in, trailed by a cymbal sounding less like a delicate accent and more like an unruly child banging pot lids. Two young women begin to sing, but they maintain only a weak allegiance to the guitar chords, letting the melody pursue a harmonic logic all its own. The singing itself is featureless and flat, displaying not the faintest trace of pop-star posturing. The song's key—the acoustic equivalent of a painter's color palette—changes in odd places. And amid all this discordant strangeness are lyrics so simple and childlike that it is impossible to tell if they represent the pedestrian longings and grievances of adolescence or an ironic attempt at profundity:

I'm so happy when you're near
I'm so sad when you're away

Austin took on the role of producer that day in the studio, calling himself the band's "proprietor." During the sessions, the girls would sometimes stop in the middle of a take. Shaggs chronicler Irwin Chusid writes that the engineers would turn to the proprietor and ask, "Why'd they stop?" Austin would reply, incredulous, "Because they made a mistake!"

After hearing the girls play off-kilter and off-key, the studio staff felt pangs of guilt for emptying out the Wiggin family pockets. They were getting paid $60 an hour in 1969 dollars, or about $456 an hour today—an exorbitant sum for a mill worker with a wife and seven children. One of the engineers let Austin know what he truly thought: that Austin was throwing away his hard-earned cash, because his daughters possessed no musical skill whatsoever. Austin listened politely but recalled his mother's infallible prophesizing. He knew in his heart that his daughters were destined for greatness. He ignored the engineer's opinion and encouraged the girls to finish recording their first album, *Philosophy of the World*.

What happened next is legend—both ornamented myth and undocumented oral history. Austin paid for one thousand copies of his daughters' album to be printed as vinyl records. In the usual telling, the engineer ran off with nine hundred of them—an odd and unconvincing twist, considering that everybody in the studio thought the music was worthless. Another version of the story claims that Austin deliberately left them at the studio. A studio client recalled the owner saying, "Austin refuses to sell these [remaining nine hundred] because he's afraid someone will copy their music."

What isn't disputed is that Austin distributed around one hundred copies of *Philosophy of the World* to radio stations and record labels. Nothing came of it. None of their songs were played on the air. Not a single talent scout reached out. The Shaggs were completely ignored outside of their hometown of Fremont, where the sisters continued playing weekly performances at the town hall and—as always—continued to get heckled and pelted with cups. That's because when the Fremont locals listened to the Shaggs, the music sounded the same to them as it did to the Boston recording engineers: "painful and torturous," as one town hall regular described it.

When Austin died of a heart attack in 1975 at the young age of forty-seven, the Shaggs disbanded. His obsession with fulfilling his mother's prophecy was the only glue holding the band together, and the sisters themselves—now in their mid-twenties—couldn't wait to escape from their proprietor's domineering control. "We ended it and went on with our own lives," Helen recounted to the BBC. "That was one life and now another." The Shaggs seemed destined for the same fate as 99.9 percent of adolescent garage bands, serving as a colorful but somewhat uncomfortable footnote in the Wiggin family history—and nothing at all in the history of music.

Then in 1980, the story of the Shaggs took an unexpected turn. Terry Adams, the keyboardist in the eclectic underground rock band NRBQ, got hold of one of the Wiggin sisters' records. When he listened to the Shaggs, he heard something quite different than the Boston recording engineers, commercial radio stations, and hometown folks of Fremont had. "Their music has its own structure, its own inner logic," Adams said admiringly. Believing he had found a diamond in the rough, he managed to convince Rounder Records to reissue *Philosophy of the World* in 1980. For the first time, a broad swath of music professionals listened to the Wiggin sisters . . . and many registered the same astonished appreciation as Adams. Avant-garde rock pioneer Frank Zappa declared, "They're better than the Beatles—even today." *Rolling Stone* magazine called the re-released album "priceless and timeless" and named the Shaggs the "Comeback of the Year." And the entertainment industry's voice of cool, the *L.A. Weekly*, archly observed, "If we can judge music on the basis of its honesty, originality, and impact, then the Shaggs' *Philosophy of the World* is the greatest record ever recorded in the history of the universe."

Chusid quotes Lester Bangs, the legendary rock critic portrayed by Philip Seymour Hoffman in *Almost Famous*, describing the Shaggs' charm: "They recorded an album up in New England that can stand, I think, as one of the landmarks of rock 'n' roll history. . . . They can't play a lick! But mainly they got the right attitude, which is all rock 'n' roll's ever been about from day one."

I first listened to the Shaggs in the late 1980s, after hearing them praised by a colleague as one of those bands that professionals love but the public ignores—or, in the Shaggs' case, actively spurns. I listened with amazement and instantly understood why they stirred up so much controversy.

Go ahead and listen to "I'm So Happy When You're Near."

What, exactly, did industry pros hear in this music that so many others had apparently missed? What made us believe we were listening to something worthy of attention?

3

Nobody was suggesting that the Wiggin sisters possessed unrecognized musical aptitude. No, from a technical point of view, they truly were bereft of musical talent—yet their music originated from the same place as every great artist's. The Shaggs' music reveals the simple yet deeply entrenched human desire to express oneself. The very fact that they failed to master formal technique renders this desire vivid and bare. They are the musical equivalent of a child's first drawing of Mom and Dad: Dad's legs are half the length of his arms and Mom appears to have three eyes, but the purity of intention is unmistakable. What my colleagues and I heard when we listened to the Shaggs was the distillation of music's lifeblood: *authenticity*.

Authenticity is the subjective conviction that the emotion expressed in a musical performance is genuine and uncontrived. Authenticity reflects a meaningful thread in the tapestry of human experience, whether a nuanced sentiment like pathos or bemusement or a strong, basic emotion like joy, fear, or sadness. Any record producer worth her salt is always looking to capture emotional authenticity in a recording, because the frank expression of feeling is a primary channel through which music connects with listeners. Musical *technique* must be learned. But musical *feeling* is instinctive, and thus readily appreciated.

You can hear authenticity in early-twentieth-century recordings that were made before performers worried about record deals and music videos. You hear it in the performance of a musician who thinks no one is listening. You see it in the finger paintings of a child and taste it in the cookies at a local bake sale. When the act of creation itself is the end goal, rather than adulation or monetary gain, the quality of the outcome may not be assured—but the *intention* can often be more readily felt.

Like other dimensions of your listener profile, authenticity itself is not easily quantified. A record's perceived authenticity lies solely in the mind of the beholder. If you are touched by a song, if you feel that the performers believe what they're playing and singing, then no one can argue that your reaction is illegitimate or misguided. Nevertheless, music scholars have provided us with a useful analytical tool to evaluate *how* authenticity is expressed, on a continuum ranging between two opposite poles.

The Shaggs lie at one extreme on the dimension of authenticity. Their music is termed "naïve": art born of no formal training or unsullied by pretension, vanity, artifice, or concern with musical rules and theories. They are describing, on their own guileless terms, exactly what it feels like to be a teenage girl in a small town. They sing about their family cat ("My pal Foot Foot / Always likes to roam"), they sing about Halloween ("Even Dracula will be there"), they sing about daydreaming and curiosity ("I wonder why my mind drifts astray?"). As Cub Koda, the leader of the band Brownsville Station, put it, "There is an innocence to these songs and their performances that's both charming and unsettling." Charming due to its easily recognized sincerity, but unsettling because it is socially risky to be that publicly naïve.

I've produced two albums by the band Geggy Tah, whose lead singer, the accomplished songwriter Tommy Jordan, taught me a great deal about authenticity. Tommy calls naïve music "music from the neck down." These compositions and performances express emotions that seem to bypass the circuits that restrain our social behaviors, delivering music that sounds as though it comes straight from the heart, guts, or hips. The naïve, below-the-neck authenticity of the Shaggs reminds record makers of what honest, uncorrupted feeling sounds like. I've listened for it in every record I've made since I first heard *Philosophy of the World*.

The opposite of naïve music is sometimes called "cerebral" music. Composers and performers of this kind of music express their feelings using deliberate principles and well-honed craftsmanship. Johann Sebastian Bach is a good example. His music communicates a wide array of potent emotions, from dramatic expressions of triumph and sadness to more nuanced feelings of longing and spirituality. He accomplishes this feat not by spontaneously expressing the tides of his heart but by carefully deploying a well-honed arsenal of polished techniques. Simply put, Bach could authentically express sadness without being sad. Musically untrained listeners can experience the sadness (or joy or anger) of Bach's music in an immediate and intimate way, while a musically trained listener can deconstruct Bach's methods and identify the specific compositional techniques he used to achieve his emotional effects.

To be clear: *Bach's music is no less authentic than the Shaggs'*, but unlike theirs, it was mindfully constructed using an elaborate scaffold of music theory and calculated rules of engagement.

Tommy Jordan calls cerebral music "music from the neck up"—a product of the brain rather than the hips, chest, or groin. Though true maestros like Bach can evoke a dazzling array of feelings using a formal system of rules, less talented musicians working from the same set of rules sometimes produce music that sounds stiff, self-conscious, or soulless. In their effort to achieve perfection, some performers take great pains to avoid below-the-neck impulses that, in the grand scheme of things, give music a human and appealing quality. When their songs or performances lack authenticity, cerebral composers and musicians can sound to listeners like they are *thinking* rather than *feeling*. If your record is technically competent yet prompts a national music critic to write, "Not only are the lyrics sadly undistinguished, but much of the production and arrangement is unusually self-indulgent and cluttered with effects"—then it could have used a generous dose of hips and heart. James Brown called this sound "Talkin' Loud and Sayin' Nothing."

Authenticity is a highly influential dimension of your listener profile. It offers an enigmatic but potent reward: the experience of genuine emotional truth. Some of us listen for it in the exposed raw nerve of gut-bucket blues, while others listen for it in the intricate elegance of compositional genius. Though I greatly prefer music with an obvious below-the-neck feel to it, many listeners are partial to above-the-neck authenticity, including my coauthor. Listen to a minute or so of one of his favorite pieces of music, a Bach composition known as Magnificat in D Major (BMV 243). In it, Bach expresses a feeling of transcendent exultation that exhilarates Ogi. The emotional connection Bach achieves flows from five-part counterpoint, symmetric and perfectly balanced intervals. Even listeners unfamiliar with formal musical technique can feel the piece communing directly with their soul.

Beyond their fresh authenticity, another reason that the Shaggs prompt many professionals, including myself, to consider them a vital musical touchstone is because they managed to communicate their emotional truths in a unified and distinctive manner. These young women, locked up in a rural home by an uneducated and oppressive father, devoid of any of the lessons provided by healthy social experiences or interactions with the wider world (let alone formal musical training), nevertheless experienced the same adolescent yearnings every teen does. Despite—not thanks to—Austin's controlling hand, they transformed those feelings into a characteristic sound unlike anything else. Individually, each Wiggin sister was musically inept, yet the Shaggs expressed themselves as one. While the sisters' individual performances would sound out of place in a practiced musical ensemble, together the Shaggs' drumming, singing, and guitar playing becomes an unexpectedly satisfying whole.

To a record maker's ears, this is the Shaggs' most intriguing quality. Chusid describes how the engineer who recorded *Philosophy of the*

World observed the Shaggs stopping in mid-performance to correct one another, saying, "No, it should sound like this," demonstrating that they were all tuned in to the same cryptic target. The legendary record producer Tony Berg admiringly raised the question "How is it possible for three people to be so perfectly wrong together?" while the blues singer Bonnie Raitt affectionately declared, "The Shaggs are like castaways on their own musical island." And like Darwin's finches on the remote Galápagos Islands who evolved their own unique birdsong, the Shaggs evolved their own musical language.

Though countless musicians have performed Bach's compositions, each putting their own spin on Bach's voice, it's futile to do a cover of a Shaggs song (although some brave souls have tried). The Wiggin sisters' melodies and lyrics are not especially interesting or compelling. It's the way they *perform together* that's so special and endows the Shaggs with their (for me) beguiling below-the-neck authenticity.

Like Emily Dickinson, the Wiggin sisters fashioned a collaborative poetry born out of isolation that followed its own peculiar rules and diction, converting their mutual loneliness into something beautiful and transcendent. Dickinson herself was a fan of naïve authenticity. The Amherst poet famously wrote, "Nature is a haunted house, but art is a house that tries to be haunted." What she meant was that, for her, self-conscious attempts at describing an interpretation of truth are never as appealing as the natural expression of those same truths.

Let me hasten to add that if you listened to the Shaggs and decided that, regardless of any educational merit their music might possess, you simply *don't like them*, please don't worry. The point of this chapter (and this book) is not to proselytize about what good taste in music should sound like but, rather, to help you better understand the sound of authenticity that speaks most compellingly to you. If you found yourself more inspired by the Baroque sophistication of Bach than the eccentric earnestness of the Shaggs (or if you responded to neither record—or both), then you learned something useful about your appetite for authenticity and where you like expressivity to come from in your favorite records: above or below the neck. Your sweet spot on the dimension of authenticity reflects your personal ideal of what constitutes heartfelt emotional expression.

This leads to a natural question, the question that drives this book: What is it about *you* that makes you feel the thrill of resonance when you hear one record but the chill of apathy when you hear another? What is going on in my coauthor's brain that makes him respond to the precise intellectual craft of Bach, while I prefer music that expresses the awkward imperfections of a performer's heart?

More simply, what makes a person fall in love with a record?

4

After I reached the peak of my career as a producer, making records that connected with millions of people, I was still curious about the wild diversity of the human response to music. Knowing the inner workings of hit (and miss) records only served to make this diversity more mysterious. I began to wonder whether the science of mind could advance my understanding of music and illuminate why music had meant so much to me throughout my life. So, in my mid-forties, I decided to take my leave of the music industry and enroll as a college freshman.

The last time I had attended school was my senior year of high school, right before I dropped out to get married. Going in, I was anxious about being the oldest student in my class at the University of Minnesota, but I felt right at home from the start. I discovered that my middle-aged brain was ready and able to learn an entirely new discipline. I graduated with a dual major in experimental psychology and neuroscience. Next, I headed farther north to study at McGill University in Montreal, earning my PhD under the tutelage of Daniel Levitin (who graciously let me read early drafts of *This Is Your Brain on Music*) and the world-renowned psychoacoustician Stephen McAdams. Now I'm a professor of psychoacoustics and record production at Berklee College of Music. I have my own music research laboratory.

After twenty-two years pursuing hit records in the studio and nearly as many years studying music psychology as a scholar and scientist, I've come to believe that the best way to understand why you fall in love with a record is by understanding your listener profile.

Your listener profile comprises seven dimensions of music that, taken together, reveal why the rewards you experience from music listening are unique to you. The sweet spots on your profile were formed through biology, experience, and happenstance. Your random neural wiring, your exposure to musical culture in the time and place you grew up, and the sheer chance of hearing *this* record and not *that* record at crucial moments in your life all shaped the kind of listener you are and influence the kind of music you can fall in love with.

For generations, scientists presumed that there was a blueprint for a "normal brain" hidden within our species' DNA that caused all individual brains to be variants of this "standard template." Today, we realize that this is not true at all. Every part of your brain follows its own unique and unpredictable developmental trajectory, leading to the formation of neural circuitry in your head that is unlike anyone else's. Because your brain is wired to experience rewards from different facets of music than my brain, it is misguided to suggest that anyone's taste in music is superior to anyone else's.

I greatly prefer the Rolling Stones to the Beatles, for example. The blues speak to me in a more compelling way than pop music and have since I was a child. This wasn't something I cultivated. It was something I *recognized*. I still remember the Rolling Stones performing "Time Is on My Side" on *The Ed Sullivan Show* when I was seven. I was astonished at the intensity of my reaction. The performance lit me up. This was *my* music! My reaction to the Beatles on *Ed Sullivan* eight months earlier, on the other hand, had been far more analytical. I peered quizzically at the television set, searching for a clue as to why all the neighborhood kids were so crazy about them.

Today I realize that the Rolling Stones hit closer to my authenticity sweet spot than the Beatles do. With their impulsive, blues-based sound, the Rolling Stones captivate me in a way that's similar to the Shaggs. The more carefully crafted artistry of the Beatles, on the other hand, delivers colossal rewards to millions of listeners, just not to me. This has no bearing on whether one band is superior to the other. I feel great admiration for the Beatles' masterful musicianship . . . but not the thunderbolt of love. (My coauthor, on the other hand, warms to the above-the-neck perfectionism of the Beatles more than the coarser, looser energy of the Stones.)

The unique developmental pathways of our neural circuitry account for much of the variety in what humans find rewarding in music, yet the mere physiological mechanisms of hearing can't explain why we feel such an intensely personal bond to music to begin with. Fortunately, neuroscience research has begun teasing out something amazing about the human brain that not only helps us understand why different listeners fall in love with different records, it explains why we feel such an intensely personal connection to our favorite music at all.

Our most private dreams and fantasies—the fears, hopes, and longings nestled deep in our psychic core—are all bound together in a newly discovered neural network in the brain, one associated with our sense of self. The discovery of this unknown brain structure unveiled an even bigger surprise: it turns out that one of the best ways to activate this personalized "self network" is by listening to music that resonates with the sweet spots on your listener profile.

The music you respond to most powerfully can reveal those parts of yourself that are the most "you"—those places your mind unerringly returns to when it is daydreaming or fantasizing. Thus, by learning about the qualities of music that match up best with your listener profile, you will not only become a better listener, you'll become better acquainted with your innermost nature. Perhaps one reason we prize our own notion of musical authenticity is because our conscious *experience* of authenticity is rooted in the brain network that embodies our self-image.

Be it records or romantic partners, we fall in love with the ones who make us feel like our best and truest self.

Musical Anhedonia

When you are enjoying music, your brain's music-listening networks communicate with your dopamine-reward network to deliver a personalized sonic treat. For listeners whose brains have diminished connections between their reward network and their musical networks, the outcome can be musical anhedonia: an absence of pleasurable responses to music.

Musical anhedonia affects an estimated 5 to 10 percent of the population. Individuals with this condition generate a normal amount of dopamine activity in response to art, food, money, and other types of stimuli—just not to music.

CHAPTER 2

REALISM

This Is What Music Looks Like

◈

She accomplishes that long-sought-after thing:
a true reciprocity between image and abstraction,
one emerging out of the other.

—David Salle, *artist, on the modern
action painter Amy Sillman*

1

LET'S KEEP ROLLING WITH OUR RECORD PULL. NEXT UP are two very different tracks. As you listen to each one, close your eyes and focus your attention on one thing: the *imagery* (if any) that appears inside your mind. Don't worry about whether you like the records or not—the point of this exercise is to consider the question, What do I visualize when I listen to music?

The first record is "Born on the Bayou" by Creedence Clearwater Revival.

The second is "The Grid" by Daft Punk.

Don't read any further until you've listened to a minute or two of each track.

Done? Good.

As you listened to each record, what did you see in your mind's eye? Maybe you envisioned a *story* unfolding that featured yourself, the vocalist, or make-believe characters. According to research I conducted with my coauthor, about 19 percent of people picture a story based on the lyrics when they listen to their favorite music. Or did you picture musicians performing the song—a band playing onstage, in a studio, or in a video? Around 17 percent of people visualize the performers. Or did you imagine that *you* were singing or performing the music? (About 11 percent of listeners.) Perhaps scenery unrelated to the lyrics took shape in your mind—a river, a mountain, a planet? (About 3 percent.) Or things you would like to build or create? (About 6 percent.) You may have seen imaginary worlds, such as in science-fiction movies (about 9 percent). Or did you see patterns of colors or shapes that didn't represent anything specific? Just over 1 percent of music listeners see abstract shapes and colors. If "Born on the Bayou" or "The Grid" happened to be familiar to you, hearing them again may have prompted you to recollect scenes from a time in your life when you listened to them. Indeed, the most common form of visualization while listening to music is autobiographical memories (25 percent of listeners).

Or maybe you are a member of the roughly 9 percent of the population who don't see *any* mental images while listening to music?

This chapter is about what you see when you listen to music, but more important, it's about what your mind *wants* to see—those private landscapes where music takes you. As we learned with authenticity, the musical experiences you unconsciously (or consciously) seek can reveal the kind of rewards your brain craves. In this chapter, we will explore the different ways listeners "see" music and what your own go-to imagery suggests about you.

The type of visualizations that naturally form inside your mind's eye when you listen to music constitute another dimension of your listener profile: *realism*. Some works of sculpture, cuisine, poetry, and painting resemble observable reality. They lead most people to arrive at the same "canonic" interpretation of what the work represents. The *Mona Lisa* is a good example. Any viewer can agree that Leonardo da Vinci's painting depicts a young woman with folded hands, dark hair, and an enigmatic smile.

Other works have a looser affiliation with reality. They illustrate the pole opposite from realism: *abstraction*. Abstract works evoke subjective and highly individualized interpretations. Franz Kafka's novella *The Metamorphosis*, for instance, tells the story of a salesman who wakes up one morning to find that he has turned into a giant insect, without revealing why this happened or what, if anything, it symbolizes. There is no canonic interpretation of this peculiar tale because even scholars cannot agree on what a salesman transforming into a bug is supposed to represent.

Similarly, some records are more realistic, while others are more abstract. The emergence of abstract records might be the most significant revolution in music since Thomas Edison invented the phonograph in 1877.

2

Abstraction in music is a very recent development, though this may come as a surprise to anyone under the age of forty. These days, abstract records—records that mostly or entirely feature computer-controlled, machine-based sounds rather than humans playing acoustic (including electronically amplified) instruments—have rapidly come to dominate the global musical soundscape. By this definition, nearly all the *Billboard* No. 1 hits in 2021 were highly abstract. The only exceptions were Adele's "Easy on Me," with its sparse piano and kick drum arrangement, and Mariah Carey's enduring holiday favorite "All I Want for Christmas Is You," recorded in 1994, when realism still dominated the airwaves. The acoustic guitar in Lil Nas X's "MONTERO (Call Me by Your Name)" and the piano in Olivia Rodrigo's "drivers license" are typical of today's abstract records: a lone traditional acoustic instrument is enveloped by sampled and machine-generated sounds.

For most of human history, all music that listeners heard was exclusively realistic. Before Edison assembled his favorite invention—a stylus that carved a soundwave into the soft wax of a rotating cylinder—music consisted solely of live performances: real people playing real instruments in real time. Audiences not only listened, they *watched* as an orchestra, parlor quartet, or jug band played and sang. So when a device appeared that enabled listeners to hear music on demand in their own home without needing to visit a concert hall, recording professionals began their century-long pursuit of making records that sounded like a live show.

Throughout almost the entirety of the twentieth century, record-ing professionals nurtured one obsession above all others: *high fidel-ity*. When I was coming up in the music industry in the 1970s and '80s, the engineer's craft still consisted of selecting the right equip-ment, materials, and techniques to re-create a musical performance so faithfully that the listener could imagine that she was sitting directly in front of the band.

Realism on a record is conveyed primarily by the *type of sounds* we hear (acoustic versus virtual instruments) and the *fidelity*, or exact-ness, with which they were captured (high or low fidelity). But there is a third factor that influences our perception of the realism of a record: the musicians' *performance gestures*. These are the unique ways an artist uses their voice or instrument to express their musical ideas and emotions. Some performance gestures become iconic, like actor Robert De Niro using a clenched, downturned frown and tilted chin to convey "Yeah, maybe" or Meryl Streep using pursed lips to mean "Hang on, I'm thinking . . ." Some musical performance gestures are so distinctive that they become an artist's "signature sound," such as the nineteenth-century violinist Niccolò Paganini's use of pizzicato (string plucking), the early-twentieth-century "cow-boy yodeling" of country music pioneer Jimmie Rodgers, Maybelle Carter's "Carter scratch" guitar technique, and Jerry Lee Lewis's pounding rock 'n' roll piano.

Performance gestures can be subtle yet offer powerful clues as to what the performer was feeling. We are intuitively aware of how bod-ies and voices convey emotional and even physical states, and these can be expressed through delicate acoustic "shadings," such as the character of the breath a singer takes between phrases or the speed at which the drummer pushes the hi-hat to ramp up to a chorus. Consequently, for most of the twentieth century recording engineers focused on capturing every subtlety of real-time performance ges-tures to make the recording feel as realistic and as characteristic as possible. By meticulously rendering musician-specific acoustic idio-syncrasies, recording professionals tried to draw you in and let you hear—and visualize—the performance as if you were right there in the studio.

The Rolling Stones' "Jumpin' Jack Flash" and Regina Spektor's "Eet" are good examples of realistic records with conspicuous perfor-mance gestures. We can easily picture Keith Richards barely making it to the mic in time for his backing vocals. It is not hard to imagine Regina caressing the piano keys as she sings.

Records that feature the high-fidelity sound of real (not virtual) instruments and preserve the musicians' unique performance ges-tures are considered "realistic records." (The Audio Engineering Society formally labels them "Traditional Acoustic Recordings.") If you've listened to music recorded in the previous century, then you've experienced the naturalistic achievement of a realistic record.

Listeners who prefer realistic records usually enjoy imagining the actual musicians performing the song—or *themselves* perform-ing the song. Because realistic music is played at a "human" speed with "human" melodies, it is easy to sing along with the vocals or pretend that we are playing one of the instruments or conducting the orchestra. For those of us whose brains are partial to realistic records, the listening experience can mirror what it's like to play or sing and, for many of us, that can be a powerful reward.

Realistic records target my own sweet spot because they enable me to experience the kinds of musical fantasies I find most grati-fying: images grounded in the human performance of the piece. I automatically *see* the musicians playing or imagine myself as one of them. I love virtuoso jazz pianist Bud Powell in part because I enjoy pretending that my own fingers can fly across the piano keys with his unparalleled dexterity. I can't resist dancing my fingers in the air to mimic what I believe he's doing, though I have no clue how to play the piano. For those of us with a sweet spot for realistic-sounding music, visualizing ourselves performing matches the satisfaction others might feel smashing a forehand like Roger Federer or owning the catwalk like Gigi Hadid.

A different kind of realistic fantasy takes shape in my mind when I hear the ethereal and melancholic voice of Lana Del Rey, one of my favorite artists. I like to imagine that I am her collaborator, sitting with her behind the console in the studio, going over the details of lyrics and arrangements. When I listen to Led Zeppelin, on the other hand, I usually imagine myself being part of the audience, watching the band from a front row seat. But no matter the performer, when I listen to music my mind immediately tries to construct a visual fantasy anchored in the real world. Thus my favorite records are the traditional kind: blues, rock, jazz, and soul, perhaps because their realism lets me put myself in the musical scene.

This brings us back to Creedence Clearwater Revival and "Born on the Bayou." The record is a fairly conventional example of late-1960s American rock, influenced by the contemporaneous trend toward longer "album cuts" that would come to dominate the 1970s and be suited for the FM radio stations that played them. I don't think that "Born on the Bayou" is exceptional as a *song*, but I do think it is a very good *record*. It builds slowly, the guitar setting a mood of suspense until we hear congas come in with the drums and so we know that this groove (that is, the rhythm) will be emphasized. The rhythm shifts at the verse and gets more staccato, more urgent. The "weight" of the record's motion sinks lower, into our hips and knees. And then that vocal! John Fogerty *attacks* the microphone from the very first line, telling us that when he was a little boy his papa warned him that the man would get him and "do what he done to me." The vocal performance lets us hear masterful and authentic technique. Fogerty deliberately controls the release of air on each line (only the best singers have such breath control). He commits 100 percent to this performance, singing with potent emotional reserve and lung power. We can *feel* the bayou's swampy presence—its lethargic, sinister passion—even if we've never traveled south of the Mason-Dixon Line.

Listen to Fogerty as he practically pants the second verse about his hound dog chasing down "a hoodoo there." I am right there with that dog in my mind's eye, and it feels so good! By the time Fogerty is wishing he were "rollin' with some Cajun queen" on a fast freight train a-choogling down to New Orleans, I want to be that woman and go "a-choogling"—a made-up word that perfectly fits the earthy spirit of this record.

Fogerty's passionate vocal performance brings the whole record home for me. "Born on the Bayou" ignites the realistic fantasies I crave, while letting me feel the below-the-neck authenticity I savor.

3

Achieving realism with magnetic tape recording was no easy feat. True masters of the art of high fidelity have always been few and far between. What separated the good engineers from the great ones was *craftsmanship*: the painstaking execution of hard-won tricks of the trade. High-fidelity techniques included, for example, the balanced capture of all forty-seven strings of a harp with just one microphone, as well as more creative tricks, such as placing a mic at the end of a garden hose with its other end in front of the kick drum or putting a mic and a loudspeaker in a cement stairwell for some do-it-yourself reverb.

One of the masters of high fidelity I worship is Mick Guzauski. He is a meticulous engineer who excels at microphone technique and the terrifically difficult art of mixing (combining the individually recorded tracks of instruments and vocals into a final stereo blend). I still remember the day in 1983 when I visited the Westlake studio where Mick was working. He had just finished mixing

Billy Idol's song "Catch My Fall." My boyfriend was Westlake's studio tech, and so I was allowed to enter the control room, a highly unusual occurrence—recording sessions at most studios are private. (Prince called his own studio policy "no noncombatants.") As Mick went to get a cup of coffee, he nodded at the tape machine. "Go ahead and play it if you want." It was an invitation from a deity to hear the angels sing! I pressed Play on the tape machine. Oh! The crystal-clear fidelity of that mix! It was beyond anything I had ever heard before, and I remember thinking, *I can tell what color socks the drummer is wearing!*

As an exceptional mix engineer, Mick makes records that can transport listeners fully into vivid mental imagery, the same way great TV shows and novels erase the veil of artifice and immerse you in a world of soldiers, queens, or astronauts.

The reason that capturing acoustic reality with magnetic tape is more difficult than it looks is because no matter how carefully you sculpt the sound going into a tape machine, it sounds different when it comes back out. Each machine and each brand of tape has its own response curve, determining how much your recorded sounds will change when they are played back. If tape made a perfect copy of sound, we would say it has a *linear* response curve. In reality, magnetic tape does an imperfect job of recording. This mismatch between the input and the output is why the response curve for tape is *nonlinear*. Engineers in the high-fidelity era developed an ear for tape's response curves, similar to how painters learned the way a paint's color changes as it dries.

Perhaps the most impressive and elite example of realism during the high-fidelity era—the recording equivalent of painting a Vatican fresco—was the direct-to-disc technique. It's exceedingly rare today, though it was unusual even during the heyday of high fidelity, because it demands grandmaster talent. When recording direct to disc, you don't use tape at all. You cut the song directly onto the master disc that will be used to press vinyl records. A cutting lathe is brought into the control room, and its stylus, or "cutting head," is fed directly from the stereo output of the recording console. The direct-to-disc method eliminates tape's idiosyncrasies and artifacts, and gives engineers the potential for phenomenal high fidelity.

Direct-to-disc recording is an Olympic-level feat because once you start the performance, you can't stop. The musicians need to play the songs in the exact order in which they'll appear on the record, with just the right amount of space between each song. Direct-to-disc recording condenses all the varied stages of record making into a single performance, much like producing a live television show. Every instrument and voice must be perfectly in tune and in sync, and no one can miss a note. It is not possible to overdub instruments or vocals later. On top of that, the mic placement, equalization, reverbs, compressors, limiters, and blend of instruments must also be perfectly adjusted in real time during each song's performance. There is no chance to edit an unwanted noise or fix an out-of-place note, other than throwing out the lacquer and starting over again from the top.

One of my favorite direct-to-disc recordings was a Sheffield Lab release titled *I've Got the Music in Me* by Thelma Houston and Pressure Cooker, recorded by engineer Bill Schnee. How I studied that record for its sublime sound quality! Even as a professional, I realized that unless I recorded direct-to-disc myself, I had no hope of attaining Schnee's near-perfect sound. Though I eventually acquired the ability to consistently reach high fidelity, I never got a chance to take my shot at direct-to-disc. As it turned out, I wouldn't need to. In the mid-1990s, just as I achieved mastery of my engineering and mixing skills, a new technology came along that would reshape the nature of music and launch the revolution in musical abstraction.

This game-changing device was known as a DAW.

4

Before we can appreciate what a DAW does and its momentous impact on music, we must first consider a similar technology-driven revolution in the visual arts. Almost two hundred years ago, a new machine appeared that changed painting forever. The artist whose story perhaps best mirrors the experience of those of us who lived through the DAW revolution in music was an early-nineteenth-century landscape painter named J.M.W. Turner.

Turner had a reputation for brusqueness and impressive skill at creating lifelike scenes with oil colors. His realistic approach to painting stood firmly within the dominant tradition of high fidelity in the European visual arts, a tradition dating back to the fourteenth century. Ever since the Renaissance, Western artists had ventured to create portraits, landscapes, and still lifes that faithfully reproduced how the eyes and brain viewed the world around them. Leonardo's *Mona Lisa* and Michelangelo's *The Last Judgment* are considered classics in no small part because of their uncanny verisimilitude. One of the most celebrated European masters of realism was Rembrandt, who deployed virtuoso lighting effects to depict facial emotion so convincingly that his figures seem to clamber out of the canvas.

No wonder, then, that artistic techniques used to convey realism, such as perspective, foreshortening, modeling, and chiaroscuro, were highly prized and closely guarded by the painters who mastered them. During the centuries-long era of realism, *craft*—skillful technique—was the primary focus for professional artists. Aspiring painters sought out mentors who would share their tricks, just as apprentice carpenters and stoneworkers did. Painters could be inventive during Rembrandt's time, of course—as long as they painted easily recognizable objects or scenes. Rembrandt was a master artist, but first and foremost he was a master craftsman. The most venerated painters of the seventeenth and eighteenth centuries were celebrated for their ability to proficiently smear dabs of colored oil upon stretched canvas so that they took on the convincing appearance of wrinkled cloth, weathered wood, decaying fruit, iridescent glass, or cherubic flesh.

As a young man, J.M.W. Turner was also a dedicated crafts-man, specializing in authentic depictions of seascapes and scenes of marine life. In 1803, Turner hauled his paints and easel to a pier jutting into the Strait of Dover off the northern coast of France and portrayed what he observed as accurately as possible in *Calais Pier: An English Packet Arriving.*

In his meticulous work, we can recognize the crashing waves, the coiled ropes, the snapping canvas of the sails, and the flapping kerchiefs of the women on the spray-swept dock. Turner painted hundreds of similar realistic oil and watercolor paintings before his thirtieth birthday. Then, seemingly out of nowhere, a new technol-ogy appeared that disrupted Turner's career path and transformed the art of painting forever.

Photography.

With the invention of cameras and photosensitive silver-coated plates, it became possible to produce images of such extraordinary fidelity that no human painter could hope to match them. This technological breakthrough instigated a profound professional and psychological crisis for visual artists. Why hire a painter to do a fam-ily portrait if you could obtain a far better image with a wooden box for a fraction of the cost and time? And why would an artist bother trying to make the most realistic representation of a scene if he knew that he would never come close to the realism in a photo? Although the camera was initially seen as a feat of engineering, by the 1830s many art critics believed that photography had become the epitome of realism in the arts.

While the advent of photography caused some painters to ques-tion their aims, it also opened a gateway to a new creative universe. If you couldn't paint the world as it was, then you were free to paint the

Calais Pier: An English Packet Arriving, J.M.W. Turner (1803)

world as it *could* be—or even paint things that had no resemblance to the physical world at all. Turner became one of the first pioneers to embrace the challenge of developing a nonrealistic style of painting that used form and color to express the beholder's *inner* experience. He devised new brushstrokes and palettes, as well as unprecedented methods for igniting creativity. It was rumored that to accurately portray the feeling of being pummeled by ferocious weather, Turner stuck his head out of a train speeding through a storm and tied himself to the mast of a tempest-battered ship. Though modern scholars question the veracity of these stories, they don't deny that Turner immersed himself in intense sensations to invigorate his painting. (Such efforts would be characterized as "naïve" approaches to painting. Turner was trying to escape from his formal training and get "below the neck.")

In 1842, after realizing that the new technology would reform the visual arts, Turner returned to the Calais pier, but this time, instead of striving to faithfully reproduce what he *saw*, he attempted to paint

Snow Storm: Steam-Boat off a Harbour's Mouth (1842)

what he *felt*—the highly personal emotions the scene evoked in him. The painting, entitled *Snow Storm: Steam-Boat off a Harbour's Mouth*, reduces the sea and the sky to a dynamic swirl of light and shadow. His sweeping, blending strokes convey the turbulent experience of a storm at sea. It is considered one of the earliest works of abstract art.

Today, it's difficult for us to appreciate just how transformative the new abstract approach to painting truly was. We were born into a visual world drenched with abstract graphics. Billboards, Web banner ads, TV show credits, and Instagram photo effects deluge our eyeballs with a daily gusher of highly stylized imagery. But for artists of the early nineteenth century and their audiences, the prospect of viewing pictures that evoked one's subjective experience instead of objective reality was an epochal turning point.

Realistic art is about what *is*, while abstract art is about what's *left out*—and what our minds are free to fill in. In the absence of familiar visual forms, abstract art invites us to invent, imagine, and fantasize about what the painting *appears* to show, rather than what it *does* show. And as the abstractness of painting increases—as the imagery becomes more ambiguous and uncertain—the viewer becomes ever more engaged with his own personalized "story" about what is happening. As the neuroscientist Vered Aviv explains, "abstract art frees our brain from the dominance of reality, enabling the brain to flow within its inner states, create new emotional and cognitive associations, and activate brain-states that are otherwise harder to access. This process is apparently rewarding as it enables the exploration of yet undiscovered inner territories of the viewer's brain."

Representational (realistic) paintings engage areas of the brain familiar with the objects and scenes portrayed. When viewers look at portraits of people, it activates their fusiform gyrus, a part of the brain involved in face recognition. In contrast, landscape painting evokes greater activation in the parahippocampal gyrus, important for memory formation and getting to know your surroundings. Still lifes—depictions of flowers, fruit, tablecloths—generate more activity in the visual cortex than other styles of painting do, perhaps because still lifes feature known objects with which we interact.

When different styles of paintings are presented to viewers, abstract paintings evoke the least amount of localized brain activation. When you look at realistic artwork, your brain automatically identifies people, furniture, mountains, bridges, apples, and other familiar items portrayed in the painting, activating the object recognition network in your brain. But when you look at abstract art, instead of *one* crucial brain structure becoming active, *many* parts of your brain light up. After all, what's a brain to think about a late-era Jackson Pollack painting, with its figureless splashes of color? Without distinct objects to recognize, viewing abstract art demands more complex mental processing—and for some brains, that extra effort is a treat.

Abstract art's detachment from physical reality may have reached a zenith in the work of the American artist James Turrell. Some of his most famous works do not use canvas, paint, or any manipulation of physical materials. Instead, he relies solely on light itself to create simple geometric works that seem to have the weight and substance of a painting. "My work has no object, no image and no focus," Turrell explains. "With no object, no image and no focus, what are you looking at? You are looking at you looking."

Photographs fail to capture the uncanny effects of viewing one of Turrell's light creations in person. His works are, as the art critic Arthur Danto describes it, "beautiful, impalpable oblongs of luminosity that one experiences as if mystical visions." Similarly, fellow critic John McDonald writes that Turrell's works are "dull to describe but magical to experience," an opinion I wholeheartedly agree with.

I recently experienced Turrell's creations at the Massachusetts Museum of Contemporary Art (Mass MoCA), one of which consisted of an attic-sized cavity filled with a mist illuminated by pale pink light. Viewers were invited to place their heads into the pink mist. This simple-sounding setup produced a truly extraordinary, dreamlike mental effect and caused many viewers to experience pleasant hallucinations. Because there were no visual features for the brain to process—no walls, floor, or ceiling—my mind searched helplessly for a reference point before giving up and wandering about on its own terms. As I continued to look around, the pink mist eventually seemed to turn black (a perceptual phenomenon known as the Ganzfeld effect), and I found myself questioning the reality of color. For me, it was a rich and deeply contemplative experience. The two women next to me, however, giggled uncontrollably for almost ten minutes straight, utterly delighted by the experience Turrell's work evoked in their own minds and, perhaps, by the trails of light that some viewers saw as they glided their hands through the mist.

The most paradigm-shattering aspect of abstract expressionist art, epitomized in the work of Turrell, is that it is impossible to form a canonical interpretation of what the piece is about, the way we can agree that Turner's *Calais Pier* is about a boat in a storm, for instance. Each viewer quite literally sees something different in Turrell's art because each viewer's mind processes nonrepresentational work in a highly personalized way.

By changing a painter's job description from "artisan" to "visionary," photography ignited an explosion of creativity in the visual arts. DAWs did the same for music.

5

DAW stands for "digital audio workstation." Its arrival heralded the introduction of digital recording into commercial music making. Digital recording does not directly record soundwaves onto magnetic tape or a soft disc. Instead, digital recording devices (such as a computer's hard drive) receive their input after the incoming soundwave has been electronically sampled tens of thousands of times each second. This process converts the dynamically changing wave into a series of 0s and 1s, like the contrasting black and white stripes on a bar code.

Digital audio incited a revolution (and revelation!) for recording engineers. For the first time, technology endowed recording devices with a *linear* response curve. Unlike with analog tape, digital recordings featured a one-to-one correspondence between what we fed into the machine and what came back out. Simply put, digital technology made it possible to achieve *100 percent fidelity*—a pristine reproduction of a musical sound or performance.

You might think that recording professionals applauded the emergence of DAWs. After all, wasn't perfect fidelity what we were after? Instead, many of us reacted the same way nineteenth-century painters reacted to photography. We were demoralized! Virtually overnight, decades of painstaking effort spent honing the laborious craft of recording to tape was rendered obsolete and, in short order, old-fashioned. Just as some painters lamented the apparent fact that the future of visual art resided with the operators of little wooden boxes, it appeared to many of us that the future of music making belonged to the operators of little silicon boxes.

The first time I listened to a digital recording, I thought it sounded strangely boring. I wasn't alone. Digital recordings lacked the familiar harmonic distortion of analog tape, a hallmark sound that many of us had learned to love. Throughout our careers we had been boosting high frequencies on our recordings because tape gradually loses those frequencies over time. Those techniques applied to digital recording could make the music sound "bright," "harsh," or "cold." We liked the warmth of analog, we argued, the same way we appreciated the warm look of film. In analog recording, the soundwave *sinks into* the storage medium, the same way lightwaves saturate the emulsion of film stock. In analog recording, cymbals smoothly decay into the whisper of tape hiss or the soft chaos of background noise. In digital recording, cymbals abruptly plunge into the blank void of nothingness.

We were also frustrated by all the professional tasks that digital recording suddenly made easy. We had painstakingly developed a sophisticated bag of tricks to get recordings to sound realistic. Using digital technology felt like cheating. After years of refining our personal sonic signatures, it was a shock to realize that anyone could pull up a perfect kick drum or a sparkling plate reverb merely by clicking a mouse. Where was the craft in that?

Fortunately, the dark cloud of DAWs proved to have a silver lining. Even though Leonardo, Michelangelo, and Rembrandt were out, Monet, Mondrian, and Miró came flying in.

Digital recording opened entirely new vistas for what music could represent—a new world unbounded by old constraints. Just as digital video technology gives us zombies, superheroes, and fire-breathing dragons that seamlessly blend into realistic scenes, digital audio technology can construct human-sounding performances out of synthetic sounds that don't exist as real-world acoustic sources. As happened in the visual arts, the aim of musical art has shifted away from expressing human *gestures* and into the realm of expressing *ideas*—or as the art critic Ernst Gombrich put it, "symbols rather than natural signs." While this paradigm shift has reduced the realism of popular music recordings, it has ignited a revolution of creative expression. A democratizing revolution, at that. One of the first things altered by the emergence of digital technology was the budget for producing a record.

Previously, only elite recording studios possessed the necessary equipment for ultra-high fidelity, and only established musicians with major record label contracts could afford the exorbitant rates to work in those rooms. Likewise, top-grade musical instruments and A-list session players were only available to artists endowed with enough record-company cash to afford them. The limitations of our tools and personnel forced most record makers to work from the materials to the vision. Our task was to make the best record we could from whatever equipment, engineers, and performers we could get our hands on. If we wanted a Hammond B3 organ and the studio didn't have one, we rewrote the part.

Today's record makers can work from the vision to the materials. Practically any sound you can think of is at your disposal or easily created. For the cost of a laptop computer, you can amass a library of sounds and recording software with a level of fidelity that once cost thousands of dollars a day. In the twenty-first century, records compete in the marketplace not on their fidelity but on how they affect the listener's imagination.

In the digital era of music making, artistic creativity is focused on developing new forms of "abstract records," records that may not even attempt to resemble musicians performing live on known instruments. There are at least two ways to produce abstract records. The first is by using newly invented sounds. The second is by manipulating or eliminating human performance gestures.

Contemporary producers use DAWs to fabricate tones, noises, melodies, and rhythms that don't necessarily originate with people performing them on instruments. Digital sound designers can conjure up audio chimeras: synthetic waveforms that can blend the sounds of a guitar and a trombone, for instance, or a parakeet and a frog, or a man and a baby. When you listen to these digital apparitions, it is more challenging to visualize them because you are unsure of what you are hearing. And when you can't identify an object—as when viewing abstract art—your mind begins to explore highly inventive possibilities. This can deepen your immersion in the experience.

Abstract records thus foster a more personalized form of engagement because they offer the listener more room to fill in the blanks. When you listen to nonrealistic records, you become a part of the musical experience with your own private interpretation of what is playing and where and how—or by soaring away into pure fantasy. To hear an example of the old sonic world colliding with the new, listen to the work of Floating Points (neuroscientist Sam Shepherd) on his cosmic electronic EP *Kuiper*—a record that treads the waters of recognizable sounds like guitar, drums, and Rhodes piano before drifting into unrecognizable sonic eddies. This kind of creative sound design was once heard mainly in electronica and techno music, but today it has become a mainstay of Top 40 pop. The technology behind innovative sound design, such as that used on "Taki Taki (feat. Selena Gomez, Ozuna, and Cardi B)" by DJ Snake, grants emerging artists a wide frontier in which to sculpt their own one-of-a-kind sound. The deeper records delve into this frontier, where sound rather than human performance is manipulated to influence our listening experience, the more *music itself* becomes an abstraction. In the official video for "Taki Taki," we see people singing and dancing but no one playing an instrument. As with abstract art, the appeal of many modern records is in the *idea* of a traditional musical performance, rather than in its *realization*.

The second way music has become increasingly abstract is through the homogenization and "correction" of performance gestures. If Ariana Grande misses a note (rare but it happens), her producers can grab the miscued B-flat on the computer screen and turn it into a B. In the past, artists and producers would often let mistakes go by to foster realistic emotional authenticity. A good example is heard in the third verse (at the 1:40 mark) of the Crosby, Stills, Nash & Young song "4 + 20." Stephen Stills chokes on the line "I embrace the many-colored beast." He wanted to redo it, but his bandmates insisted that the mistake stay in. When I listened to this record as a young teen, the vulnerability expressed in this moment brought the image of the performer vividly to mind, which made me love it all the more.

With the speed and ease of digital editing, the practice of sanitizing recordings is no longer a pragmatic method for cleaning up objective mistakes but a creative philosophy that can motivate producers and engineers to correct subjectively perceived "flaws" in a musician's performance. Yet, by doing so, they wipe out those unique and revealing performance gestures that previous generations celebrated and many contemporary listeners still long to hear. As a result, the vast majority of today's records are technically perfect . . . but physically impossible to perform live. Or, at least, impossible to perform the way they sound on the record.

These days, my students often hear any "coloring outside the lines" as a sign of unacceptable sloppiness and a lack of effort. According to many young listeners, if it can be fixed, it should be. Consider one of the simplest performance gestures of all: breathing. Compare the wonderfully prominent inhales and exhales heard on Regina Spektor's vocal performance of "Eet" with the almost total absence of them on Ariana Grande's "7 Rings," a track that makes me tense every time I listen to it because I'm constantly thinking, *"Breathe, woman, breathe!"*

6

Just as the emergence of abstract painting revealed untapped preferences in the public's enjoyment of the visual arts, so has abstract record production. Listeners who prefer records with computer-generated sounds and performances often place a lower premium on realism than those whose taste leans toward traditional analog records. Because everything can sound perfect when using a DAW, pristine fidelity and perfect technique are not special. This vastly increases the premium listeners assign to the imaginative *vision* in a musical recording—the creativity of its sound design. As records drifted away from and eventually toppled over the cliff of realism to arrive in the valleys of sonic invention, it provoked an expansion of the ways in which a record can delight a listener, including the ways listeners visualize music.

Abstract records invite you to create your own personal world and invent your own imagery, unbounded by consensual reality. My coauthor, for instance, never imagines actual performers when he listens to records, the way I do. Unlike my own childhood experiences where I instinctively pictured what the band looked like and felt a deep connection to the imperfect and sweaty effort of making music, Ogi's youthful brain generated abstract patterns of pleasing shapes and colors when he listened to music. As a result, he initially preferred music without lyrics that relied on complex and intricate melodies, such as Bach, which seemed to produce the most intense visual patterns. As he got older, the music he found most rewarding tended to feature abstract soundscapes that invited him to imagine and wander through "impossible" fantasy worlds. A good example is Daft Punk's song "The Grid."

Daft Punk was a French duo influenced by Kraftwerk and other pioneers of musical abstraction. Guy-Manuel de Homem-Christo and Thomas Bangalter were among the first to embrace the new possibilities afforded by DAWs, experimenting with synthesizers and drum machines until they evolved a musical style all their own. Though their lush musical soundscapes are unquestionably synthetic, they stay tethered to authentic human emotion by combining computer-crafted electronic melodies with musician-performed funk and rock elements. Daft Punk embraced the new ideas and philosophies of abstract recordings so thoroughly that they created "robot personas" for themselves, always performing and appearing in public while wearing elaborate electric helmets and gloves that completely mask their identities.

"The Grid" is taken from the soundtrack for the 2010 movie *Tron: Legacy*. The album was composed by Daft Punk as a film score and combines their trademark electronic inventions with symphonic music performed by a real orchestra. Though the movie itself, about a hidden world inside a computer, was largely forgettable, the soundtrack effectively evokes the sense of an alternate universe. Instead of asking listeners to visualize what the music makers are doing, "The Grid" invites them to journey to another realm, to imagine a place that operates according to unfamiliar rules.

The record achieves its otherworldly effects by mixing familiar-sounding human-performed strings with an insistent mechanical rhythm, digitally enhanced harmonics, and a production style that relies on computer-like precision. It also features a monologue about fantasy and exploration spoken in the rich baritone of actor Jeff Bridges, disrupting the usual expectations of listening to a song and, along with the urgent arpeggios of the strings, suggesting that something dramatic is about to happen—that you are about to embark on a voyage of discovery, perhaps. For Ogi, the reward from listening to Daft Punk's record is not the pleasure of becoming one with the performer but the joy of wandering through a dreamlike reverie.

One's preference for realistic or abstract art does not say anything about one's intelligence, maturity, or level of cultural sophistication—

it merely shines a light on the highly individualized mental activities that each brain finds pleasing. Sadly, and incorrectly, the various movements of abstract art, such as impressionism and cubism, are often misinterpreted as cultural "advances" in the same way that electric cars and iPhones represent technological advances. That is, they are perceived as *superior* to what came before. Sometimes the notion is promoted that abstract art is intellectually more eminent than realistic painting. Nothing could be further from the truth. Whether you marvel at the exquisitely rendered detail in Johannes Vermeer's *Girl with a Pearl Earring* or are moved by the geometric genius of Piet Mondrian's *Composition No. 10*, it reflects the unique way your own constellation of neural networks savors what the painting evokes in you.

Similarly, before DAWs there were many listeners whose dormant tastes for "impossible" fantasies were not served by realistic records featuring of-this-world performances and sounds. Whether you favor "Born on the Bayou" or "The Grid" is not a marker indicating that you are more or less musically sophisticated; it merely tells you that your brain is wired to prefer one end of the dimension of realism over the other.

An interesting finding from the study of individuality suggests that our appetites for realism are specific to each form of art. Your visual circuits develop largely independently from your taste circuits, which develop largely independently from your auditory circuits, which develop largely independently from your olfactory circuits, which means that the rewards you experience from visual offerings can end up being quite different in style from the rewards you experience through taste, touch, smell, or sound.

This biological fact can lead to some surprising contradictions within our personalities. Even though I possess a very strong preference for realistic records, I greatly prefer abstract visual arts, such as paintings by Jean-Michel Basquiat and Cy Twombly. The joy I experience from picturing real musicians in a real studio is equaled by running my eyes over a Picasso "found object" sculpture and perceiving a bicycle seat as a bull's head. My coauthor is the exact opposite. Though he mostly listens to abstract records that enable him to travel through his own phantasmagoric psychic spaces, he dislikes most abstract art, instead preferring works where the subject is vivid, concrete, and realistic. He'd rather look at one of J.M.W. Turner's early, pre-photography seascapes than one of his later abstract paintings.

Consider your own proclivities. Do you prefer Robert Frost's down-to-earth poetry of everyday life in New England ("The way a crow / Shook down on me / The dust of snow / From a hemlock tree") or the more enigmatic work of Carl Sandburg ("And I, / The gold in the house, / Writhed into a stiff pool")? Camille Claudel's figurative sculptures (like *Perseus and the Gorgon*) or the faceless geometries of Constantin Brâncusi (like the banana-shaped *Bird in Space*)? Traditional dishes where it's possible to instantly recognize each ingredient (such as pasta with marinara sauce) or avant-garde cuisine where there's little or no connection between the appearance of the dish and its ingredients (such as "Graffiti" at Alinea in Chicago, a colorful burst of unidentifiable shapes composed of "carrot spray paint and wild mushrooms")?

The images that float through your mind while you're listening to your favorite music are linked to your core sense of self. Recognizing where your sweet spot lies on the dimension of realism clarifies why certain records are more likely to elicit the visions you seek—and prevents you from feeling like something is wrong when others prefer music that doesn't fire your imagination.

Absolute Pitch

Melodies have a curious perceptual quality. You can change all the individual pitches that make up a melody and still recognize the melody as the same, as long as all the pitches shift by the same amount. A uniform melodic shift is known as *transposition*.

Almost every human has the ability to recognize a transposed melody as the same as the original. This talent is known as *relative pitch* perception. In contrast, roughly one person in ten thousand possesses *absolute pitch*, or "perfect pitch": the ability to instantly name the note that a given pitch corresponds to, such as A-sharp or E. (Some scientists insist that to qualify as absolute pitch, its possessor must also be able to instantaneously and accurately produce any named note.)

Absolute pitch is a feat of categorization. The ability forms in childhood when neural connections link specific pitches to verbal labels, similar to how we learn the names for colors. Some individuals with absolute pitch can name as many as seventy different pitches, spanning roughly six octaves. In contrast, a listener with relative pitch could label the seven pitches used in a C-major scale (A, B, C, D, E, F, G), but only after they've managed to correctly identify the starting, or root, pitch.

CHAPTER 3

NOVELTY

This Is What Risk-Taking Sounds Like

✦

"Well, it's always been my nature to take chances
My right hand drawing back while my
left hand advances."

—*Bob Dylan, "Angelina"*

1

BRIGHT AND AFFABLE OMAR, ONE OF MY STUDENTS AT Berklee, was always surrounded by a circle of friends. He worked hard at being liked. Omar shunned the black nail polish, combat boots, and facial hair fads circulating around our urban creative-arts campus. When voicing his opinion on cultural matters, his views were more cautious and moderate than most. When it came to music, Omar pledged his allegiance entirely to pop.

For Omar, anything that wasn't riding the top of the charts was suspect. After hearing Prince's "Condition of the Heart"—to me, a sublime and unorthodox musical expression of loneliness and longing—Omar said that if he didn't know better, he would say that Prince couldn't sing. He showed no interest in venturing out onto new musical limbs to explore any forms of music other than Top 40.

Sheryl, on the other hand, exhibited zero interest in pop music. Nor did she express any curiosity for the kind of cutting-edge records that usually draw so much attention at Berklee. All her devotion was reserved for a single genre: reggae. Whenever she played her favorite reggae tracks in class, her face would light up. She often took deep dives into the histories of reggae artists and producers, a trait I appreciated because I always learned something new from her reconnaissance. Sheryl embraced this musical tradition completely, spending all her time listening to, performing, and producing reggae songs to perpetuate her beloved genre.

And then there was Andrew. His favorite records were made

by boundary-pushing artists who twisted and bent classic forms in the name of creative exploration. A natural-born record producer, Andrew set himself a goal I admired greatly: every day, Andrew listened to a new album from start to finish. Whenever I saw Andrew, I would ask, "What was yesterday's record?" and he would name a new artist, usually one I had never heard of. He always gave me a brief review of their work.

Back when I was his teacher, Andrew introduced me to American hardcore. It's a style of punk music that is more extreme (in speed and aggression) and more sophisticated (in harmonic and rhythmic complexity) than most punk. When I accompanied Andrew and a few other students to a hardcore show at the Middle East club in Cambridge, I hung out at the edge of the crowd, with my escorts huddling around me like protective squires. Just minutes into the set, lithe and slender Andrew announced, "I'm going in!" and hurled himself into the thrashing core of the most brutal mosh pit I'd ever seen.

This trio of Berklee students illuminates a listener profile dimension encoding your appetite for musical risk: the dimension of *novelty*. Humans are born with the urge to explore new objects and situations, provided we stay within our personal "Goldilocks zone" for novelty: not too strange, not too boring. Like the fabled Goldilocks trying out different beds, we seek out music that feels *just right*.

Though your appetite for novelty can vary dramatically across contexts and life stages, your personal playlist can reflect a deep-seated preference for well-traveled musical boulevards or for unmarked back alleys twisting along the edge of town. Some of us feel magnetically drawn to artists whose odd, intriguing names we heard only in passing, while others think of our music libraries as old companions we enjoy visiting and revisiting throughout our lives.

In this chapter, we explore the dimension of novelty and learn how your appetite for risk influences which records feel "just right."

2

When it comes to music, familiarity and novelty are *subjective* properties. What's musically familiar to you might be unprecedented for me, and vice versa. Provided that a listener is familiar with the music of a particular culture, it is relatively easy to classify a piece of that culture's music as standard or avant-garde. Each record is heard in relation to other records that follow the same set of musical rules. For listeners unfamiliar with a given musical system, every piece in that style will sound novel.

The melancholy, otherworldly songs of the Lebanese singer Fayrouz sound exotic to American ears, though every resident of the Middle East is well acquainted with the music of this Arab icon. Indian ragas use microtunings (musical notes that fall *between* the notes of adjacent black and white keys on a piano) that are extremely rare in Western music. Listeners well versed in ragas could easily arrange a set of them according to their degree of novelty, while listeners in the West often find that all ragas sound equally foreign.

Nevertheless, in the twenty-first century, most people share a common understanding of the basic structure and elements of Western music. To the dismay of ethnomusicologists, after pop and rock 'n' roll music originated in Britain and the United States in the 1950s, it quickly spread to become, according to the venerated *New Grove Dictionary of Music and Musicians*, the "lingua franca" of the world. For better or worse, Western forms of music now reach every corner of the globe through popular movies, TV shows, online videos, video games, advertising, and shopping mall playlists.

One element of Western music that has quietly sunk into most humans' brains without their conscious awareness is the eight-bar section. A bar is a structural unit of music. In a given musical composition, a bar has a duration specified by the composition's "time signature." The time signature defines the number of beats per bar.

Western pop music usually features four beats per bar: ONE–two–three–four, ONE–two–three–four, ONE–two–three–four. That's three bars. If you have eight such bars in a row, that's an eight-bar section.

The eight-bar section is such a ubiquitous element in modern music that it is effective at building anticipation for new events. Most listeners instinctively expect that after eight bars, the previous musical section (perhaps a verse section) will end, and a new one (perhaps a chorus) will begin.

Here are two records that are quite different stylistically, but are each built out of eight-bar sections: Steven Page's "I Can See My House from Here" and Dua Lipa's "Don't Start Now." Page's eight-bar introductory verse is followed by another eight bars of verse, then goes straight into a dynamic eight-bar chorus. In contrast, Lipa's eight-bar introductory verse is followed by a second extended verse that surprises us by pulling the plug after only four bars to take us into an eight-bar pre-chorus (a transitional section that continues to build anticipation for the chorus), which, in turn, is followed by an eight-bar chorus that delivers the release we anticipate.

Our constant exposure to modern music has taught us to anticipate something new after eight, four, or sixteen bars, whether we're listening to Broadway musicals or death metal. Music featuring even-numbered sets of bars will sound more familiar to contemporary listeners than music that changes after five, seven, or thirteen bars. The fact that most music listeners share a common if implicit understanding of the elements of contemporary music gives rise to the *novelty-popularity curve*.

The novelty-popularity curve is a visual expression of the relationship between the cultural familiarity of a record and its commercial success. The curve not only illuminates the collective dynamics of the musical marketplace; it can help you determine where your own sweet spot resides on the dimension of novelty.

The horizontal axis of the curve represents the cultural *novelty*

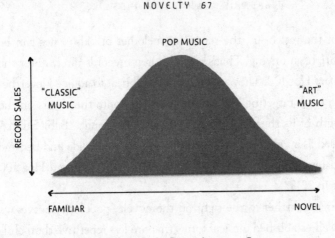

The Novelty-Popularity Curve

of a record. It ranges from the simplest, most familiar, and therefore most predictable records on the far left to the most complex, boundary-pushing, and unpredictable records on the far right. The vertical axis represents the popularity of a record in terms of its sales (including streaming). It ranges from zero sales at the bottom to the biggest global mega-hits at the top. The most commercially successful music—the *Billboard* chart-toppers preferred by my student Omar—is found at the peak of the curve, centered between the two extremes of familiarity and novelty.

At the far left of the novelty-popularity curve sits the simplest music, such as lullabies and children's songs—the type of music most listeners heard in infancy. The simplest music tends to feature uncomplicated, steady rhythms played at a moderate tempo. Its structure relies on the repetition of a few pleasant and harmonious melodic phrases, making it easy for toddlers to follow along and learn. The lyrical rhyme schemes are clear and obvious. A good example is "Twinkle, Twinkle, Little Star," which is about as uncomplicated as a viable melody can get. A song that is simple to memorize and easy to predict provides very young listeners with a fulfilling sense of participation.

For the rest of us, the repetitive melodies of kids' songs can be downright annoying. That's why children's records rarely enter the *Billboard* Hot 100. They are too banal to appeal to adults. Nevertheless, a track targeting kids occasionally slips onto the charts on the strength of its sheer mindless catchiness. Pinkfong's "Baby Shark," released as a video in 2015, fired up toddlers worldwide and became an instant children's classic, breaking onto the *Billboard* Hot 100 chart in 2019.

Moving farther to the right on the novelty-popularity curve, we find well-established musical forms that are less repetitive than children's songs and have enough variety to attract adults. I'll call these "classic" records (not to be confused with "classical music," such as composed by Mozart and Brahms). Enduring genres with highly standardized musical elements are found here, like blues, rock, country, R&B, gospel, and reggae, the genre beloved by my student Sheryl. Fans of classic records aren't seeking compositional or sonic innovation. They enjoy the reward of listening to exceptional proficiency presented in a familiar style.

"My First Lover," by Gillian Welch, is a good example of a record on the left side of the novelty-popularity curve. Welch and her musical partner David Rawlings have devoted their artistic lives to reviving and expanding upon the American roots music of Appalachia and the South. The seminal works of bluegrass, country, and Americana were written during a time of great financial hardship brought about by the First World War, the Great Depression, and drought across the plains. These classic records were made with acoustic stringed instruments that handled all rhythmic duties without the need for a drum kit. The vocals were often performed by two vocalists singing in harmony. Lyrics typically described provincial concerns of hardship and redemption through work and sacrifice. Just like rock or reggae, the canon of American roots music is well known to purists, and so to appeal to this audience, any attempts at modernization must stay true to its classic form. "My First Lover" is a familiar record in terms of its instrumentation, but it pushes into modern territory with its high-resolution sonics and four-on-the-floor pulse—the standard rhythm of disco and electronic music where the kick drum plays four evenly spaced beats per bar, instead of the usual two per bar.

Listening to Welch's record does not transport us back in time to dirt-floor cabins and barefoot children. That is not its intention. Instead, the lyrics place us squarely in the late twentieth century by referring to a Steve Miller record (a singer-songwriter popular in the 1970s and '80s). Yet Welch's vocal performance is direct and resolute, like the great performances of the past, rather than expressing the kittenish sexuality typical of pop songs in the present.

"My First Lover" and other records that advance traditional genres aim to present a controlled degree of surprise within the reliable confines of familiarity. This form of pleasure is well known to every sports fan. Basketball fans have watched hundreds or thousands of games and intimately know the rules of play, the roles of the players, the methods of scoring, and the flow of the game. Thus, every basketball game is to a large extent predictable. You know that you'll see some layups, some fouls, and some jump shots, that the game will have forty-eight minutes of playing time and a fifteen-minute halftime, and that by the end one team is going to win. What makes a new game enjoyable to a fan is the drama and suspense of reaching the outcome, which is dependent on the execution of well-established plays and strategies by extremely proficient players. If there's no basketball game on, a devoted fan won't switch over to watching épée fencing or water polo. The lack of familiarity with the conventions of these alternate sports would be a barrier to enjoyment rather than a conduit.

Similarly, listeners who prefer records on the left side of the novelty-popularity curve are rewarded by hearing meticulous execution from talented performers honoring the tried-and-true elegance of a well-established form. Fans preferring novel forms might avoid or dismiss classic records, assuming that such music won't provide the provocative twists they crave. Fans of classic records, on the other hand, savor the polished craftsmanship, which is why the songwriters, musicians, vocalists, and engineers who make classic records are among the best in the world.

3

Continuing to the right, we reach the peak of the curve. Records here represent the median level of novelty in contemporary music. This is where we find the musical style most popular with consumers: the aptly labeled "pop" music. Pop(ular) music is so called because *in the current cultural moment*, these records offer a balance of familiarity and novelty that matches the "Goldilocks zone" for most listeners' musical risk-taking appetites.

Records on the *Billboard* Hot 100 chart have a greater degree of innovation than records in classic genres, like reggae, rock, and country. Convention-busting originality in pop records is typically introduced through one or perhaps two of a record's elements, while the remaining musical elements stay in familiar territory. For example, "bad guy" by Billie Eilish and her brother, Finneas O'Connell, features a conventional rhythm and a conventional (though compelling) melodic hook. The record's novelty lies in its vocal style. Billie hugs the microphone and uses almost no projection at all. Her technique, which was uncommon when it debuted in 2019, sounded fresh and intimate and drew us in.

A pop music artist can opt to introduce novelty in a song's rhythm by leaving silences where we expect a beat or by slipping some notes off the beat. Both anticipation-violating techniques are heard in "Try Again" by Aaliyah. This record showcases the rhythmic ingenuity of producer Timbaland: it hits the listener with a tiny jolt of musical pleasure on every bar because the beats don't fall where you expect them to. Every deviation from our rhythmic predictions feels like watching a magician make a card disappear.

A pop record can incorporate innovation in its sound design by offering up familiar instruments in surprising sonic packages. "Runaway (feat. Pusha T)" by the sublimely innovative Kanye West, contains an old-school hip-hop beat along with instrumental timbres that were startling for the genre in 2010. The record starts with identical high-pitched piano notes, taking their own sweet time to get to any variation that might establish a traditional melody. The slow-paced intro violates the conventional wisdom that modern song intros should be short because contemporary listeners (supposedly) have short attention spans and won't be willing to hang around for the first verse to arrive. Yet, even with an intro this long and bare, West manages to hint that something extraordinary is afoot. We're willing to wait and see. And then, suddenly, after a full thirty-eight seconds of repetitive, unadorned piano notes, the record launches into a flood of drums, bass, and a voice shouting, "Look at ya!" Individually, each musical element is familiar. Organized in such an unorthodox way, the elements of "Runaway" collectively deliver a reward of surprise.

If your favorite kind of music is pop, your taste resides on the fulcrum where craft meets innovation, suggesting that you enjoy equal portions of these flavors on your musical plate. You're far from alone. At any given moment in time, the largest bloc of listeners will find their sweet spot at the intersection of familiarity and novelty, embodied within the songs on the *Billboard* Hot 100.

Now we swoop over the summit to the far side of the curve. Here we find records that shatter expectations more frequently than pop does. Music in this zone blazes new artistic trails. A small portion of these records will ignite new trends that will eventually become popular, while others will outpace current trends to such an extent that they earn the label "ahead of their time." Of course, some of these records will *never* have a time because they express musical ideas that most listeners simply don't like.

Music to the immediate right of the curve's peak is often called "art music," which over the years has included genres like jazz, punk, electronica, hardcore, goth, math metal, progressive rock, and grime. In time, some of these genres shifted toward the center of the curve. Others have not. Art music often exhibits a greater degree of originality or complexity than the classic genres found on the left half of the curve, yet the symmetry of the curve suggests that, collectively, art records have just as many passionate and loyal fans as classic records. Art music may feature more sections than pop (prog rock), no chorus (techno), vocals that are screamed rather than sung (hardcore), or utterly unpredictable rhythms (glitch hop). For the musically adventurous, radical music is like catnip. To the less intrepid, it is like a swarm of bees.

Art music, like art cinema, is often favored by the cognoscenti, critics, and fans who revel in taking chances. These listeners seek out new sounds at the risk of wasting their money, time, or reputation. Engaging with novel stimuli demands more cognitive effort and commitment than engaging with the routine. It can also carry social and economic costs: fans of art music spend cash and effort pursuing new musical experiences that don't always pan out, and sometimes they are mocked for their unconventional tastes. But for those with a genuine appetite for musical risk, the sheer delight of making a new discovery is worth the price.

Sometimes classic forms of music are transformed through an injection of novelty, bringing in new audiences. In the 1990s, grunge took the familiar elements of rock music and blended them with lyrics and performance techniques drawn from the right side of the curve to attract a broad following, an achievement exemplified by grunge icons Nirvana, Soundgarden, and Pearl Jam. These bands incorporated novel elements in their vocals, rhythms, and drum sounds but stayed close to traditional rock-guitar timbres and melodies. Bands like the Melvins, on the other hand, leaned more heavily toward punk, math metal, and other art music styles, and did not enjoy the same level of commercial success.

In the twenty-first century, some rock-based artists are venturing even farther to the right of the curve by experimenting with microtonality, odd time signatures, and the spoken word. You can hear all of this and more in the work of the Australian band King Gizzard & the Lizard Wizard. I recommend "Crumbling Castle" on their album *Polygondwanaland*.

Tennyson, a brother and sister duo from Edmonton, Canada, is another example of musical artists exploring unknown realms by interleaving a few familiar elements within largely novel pieces. Their 2015 single "Like What" from the EP of the same name includes the sounds of an analog telephone's dial tone and the world's easiest-to-play piano tune, "Chopsticks." These familiar features are presented against a sparse electronic background that sometimes disappears to almost nothing before reappearing in an amorphous buffet of time signatures. The resulting track is beautiful, intriguing, and inspiring for those listeners who like an ample serving of novelty.

My own appetite for musical risk-taking enriched my career because it led me to musical treasures I might otherwise have missed. Perhaps the most influential example of this was choosing to work with Tommy Jordan and Greg Kurstin of the band Geggy Tah. Their music was unlike anything I'd ever heard. Geggy Tah released their debut album, *Grand Opening*, in 1994, and critics raved about its avant-garde approach, though radio-station programmers had no use for its convention-shredding creativity. The album experiments with novel song structures, uncommon lyrical themes, inventive sonic objects, and what Tommy called "self-sampling": making new music from snippets of their previous recordings. Even though their unorthodox approaches challenged me, I was delighted by the glimmer of genius peeking through their methods.

Many facets of Geggy Tah's musical originality can be heard on the track "Ovary Z's." It's a song about a man dreaming that he is awake and experiencing a menstrual period. The very concept was risky, considering cultural sensitivities over this private bodily issue. The foundation for the song was captured by putting a mic over a turntable playing Geggy Tah's 45-rpm record "Giddy Up" at half speed. In the recording of "Ovary Z's," a sleeping man wakes and drops the needle on "Giddy Up" to remind him that "s*#t creek never catches up to the chin of those who get up." He rises, yawns, turns on the shower, slams the door, and the 45 begins to skip, repeating "giddy up, giddy up, giddy up" but at half speed—sluggish and woozy—like the feeling of waking up from a heavy slumber. With the skipping record fading into the background, the singer narrates his bizarre dream.

Even the title "Ovary Z's" incorporates layered meanings by combining the biological term for an egg ("ova") from a woman's anatomy ("ovary") with sleeping ("catching some Z's") to form the homophone "over easy"—the way many folks like their eggs in the morning. The recording is low-fidelity and lacks the polished craftsmanship you expect to hear on classic records. Geggy Tah fans were unlikely to care about that. "Ovary Z's" teems with the sort of ingenuity that satisfies listeners with an appetite for the unfamiliar.

To help my production students recognize talent as well as gauge their own appetite for musical risk-taking, I assign them the task of listening to *Grand Opening* and reporting what they glean. Sadly, these reports often include adjectives like "wacky," "goofy," or "weird." It can be disheartening for adventurous artists and producers to be so misunderstood. I must remind students that when they hear a record that sounds strange, they should consider two possibilities. One, that the artist was inexperienced or incompetent and simply didn't know how to make a commercial record. Or two, that the artist did it on purpose.

In Geggy Tah's case, the duo was extremely talented and highly trained. They knew *exactly* what they were doing. Like most artists whose work lies on the right side of the novelty-popularity curve, Tommy and Greg were striving to push music forward by exploring how far they could stretch the parameters of *what music is*.

4

Finally, we arrive at the rightmost extreme of the novelty-popularity curve. Here be dragons. This is where the most complex, novel, and *unpredictable* records lie. By conscious design, the most unpredictable musical genre of them all—aside from sonic experiments with chance, such as circuit bending—is free jazz. Artists associated with the free jazz style include Ornette Coleman, Anthony Braxton, and Pharoah Sanders. Free jazz takes the improvisation component found in all jazz music and extends it even further to create musical adventures that often sound totally unmoored from any recognizable theme or structure.

Free jazz is the antithesis of elevator music. Its rhythms tend to accelerate and decelerate. There is rarely any reliance on chord progression. It features open-ended melodic structures that are only loosely tied to the rules of tonality. On a free jazz record, the musical instruments are familiar, but practically nothing else is.

You can sample the sound of innovation in "The Sun" by the pianist Alice Coltrane, the wife of jazz legend John Coltrane. The piece is expressive and beautiful, but its ideas are far less apparent and comprehensible than those we immediately recognized on Gillian Welch's "My First Lover." "The Sun" is harder for our brains to process, but many listeners find it more compelling for exactly that reason. Free jazz has a devoted audience, but it sells far fewer records than popular styles, because the greater the novelty, the heavier the cognitive load for listeners, who must work harder to identify and follow its patterns. Fans of free jazz and other highly complex styles are motivated to put in the mental effort to track and memorize intricate phrases, and this makes them very active, engaged listeners.

Risk-taking artists play an indispensable role in the advancement of commercial music. All the artists I have worked with considered themselves to be leaning forward into original artistic expression, though constrained by the need to reward listeners. Few musicians are so naïve as to believe that their more experimental pieces will be hits, but most accept that the risk of commercial failure is the price of expanding the scope of music. Artists working at the forefront of musical experimentation who forge an appealing new idea quickly spawn imitators, inspiring new musical trends. Savvy pop producers and artists make a habit of checking out the fringe offerings. The true influencers have a talent for recognizing a good idea whose time has come.

By plucking lesser-known articles from the cultural incubator, we got kale, role-playing games, hoverboards—and hip-hop. According to the rap pioneer Carlton Ridenhour, a.k.a. Chuck D, hip-hop was invented in 1973 when Clive Campbell, a.k.a. DJ Kool Herc, threw a "back to school jam" for his little sister in the basement of their home in the Bronx. Using his turntables, Herc invented a way of looping drum sections on vinyl records to create prolonged drum breaks for dancers. During one of these drum breaks, Campbell's friend Coke La Rock grabbed a mic and began calling out friends' names and rapping improvised lyrics over the beats. From this humble origin, the rap innovation spawned imitators on both coasts and eventually gave birth to one of the most globally popular forms of music of the past half century.

5

So far, we've reviewed the *horizontal* and *vertical* dimensions of the novelty-popularity curve—the familiarity of a record's style and its commercial success. However, there is a third *temporal* dimension to the curve: how the perceived novelty of musical genres changes over time.

At any given moment, the novelty-popularity curve maintains its distinctive shape. There was a bell-shaped curve in the 2000s, in the 1980s, in the '60s, even in the '40s and '20s. But though its shape remains the same from generation to generation, the curve itself slides steadily to the right along the axis of novelty as different musical innovations become commonplace. The peak of the curve—along with the most popular style of music—retains a balance of familiar and novel elements, but *what those elements sound like* changes as audiences get accustomed to musical advances.

The novelty-popularity curve is always relative to a moment in time. A record that sounded groundbreaking twenty years ago will usually sound conventional today, especially if its style was widely imitated. New technologies, new production techniques, new rhythmic experiments, and new lyrical trends make pop music in the 2020s sound much more innovative, daring, and sophisticated than what I heard on Top 40 radio in the 1960s. To teenage listeners, today's pop music may not seem innovative at all, because they don't have much to compare it to. Older listeners are more likely to have music libraries that span many decades, and so we have a larger sample to consider when we evaluate novelty.

Record makers are constantly updating established musical norms with new instrumentation and recording methods. New timbral palettes, rhythms, and taboo-busting lyrics emerge. As novel, unproven records gain some traction and begin to demonstrate their appeal in the marketplace, they attract more and more artists willing to try the new style. This increases the style's overall complexity, because as artists experiment with a new genre, its variety grows. But once a new musical style makes its first appearance on the pop charts, the opposite trend occurs. When that happens, the style spawns more *imitators* and fewer *experimenters*. Market pressure pares down the variety in pop's timbres and stylistic choices as imitators choose to follow what they know will work.

For instance, when disco and new wave music were just starting to attract audiences in the 1970s and '80s, these records quickly climbed toward higher timbral complexity. Yet once these styles became mainstream, disco and new wave records became stylistically more homogeneous and thus more predictable, eventually finding their present place on the left side of the bell curve. New records in older genres feature less innovation to fit a now-classic form.

Because there are always plenty of listeners with an appetite for musical novelty, there will always be visionary work from risk-taking artists. Some of it will arrive at the right moment in a particular zeitgeist and gain momentum, launching the next trend. When this happens, existing musical styles that are familiar to listeners are pushed to the left of the curve. As music has evolved, many of the melodies, lyrics, rhythms, and timbres of earlier eras have grown sonically rusty. Sometimes once-beloved genres even vanish completely, as in the case of ragtime and glam metal. During the mid-1980s, I witnessed the reign of Prince, Michael Jackson, and Madonna. Their funk-inspired pop became the rage, and so their style was widely imitated. When I listen to their records today and imagine what they must sound like to kids in the 2020s, I want to blurt out, *I swear it was hip when we did it!*

Just as almost any piece of fabric can be fashioned into a wide variety of garments, so can any given song be recorded in a variety of ways. If a record maker chooses to apply the styles and techniques of current pop music, the song will stand the best chance of becoming a hit. You might surmise that success-minded record makers would follow this prescription every time, but it does involve some risk. Chasing the latest trend can shorten an artist's career by consigning them to "one-hit wonder" or "flavor of the month" status once the public's appetite for novelty inevitably drives people to seek out artists who boast an even fresher sound.

Music is an evolving creature because *what music sounds like to a generation* depends largely on what that generation listened to growing up. The pandemic year of 2020 coincided with a much-anticipated sea change in musical tastes heard at Berklee. As recently as 2018, we were still getting a few students making classic rock records, no doubt inspired by their parents' CD collections. Today, students gravitating toward classic rock have all but disappeared. The new millennium's digital revolution opened so many doors for novelty that it finally overturned what record producer Tony Berg bemoaned as a "ridiculously conservative" period in music. Grunge in the 1990s wasn't drastically different from rock in the '80s or '70s. Rap in the '90s wasn't that far removed from '80s hip-hop. But once technical innovation put affordable recording tools into the hands of every wannabe record maker, a rising tide of novel ideas saturated the marketplace. The vast majority of today's music students grew up listening to records that were aesthetically a far cry from those made in the twentieth century.

The do-it-yourself digital revolution in music production means that the novelty-popularity curve is sliding forward at an ever-faster clip. I'm in my sixties, and at long last I am finally experiencing the welcome feeling of *I don't get what the kids are listening to these days!*

6

Sheryl loves reggae, Omar loves pop, Andrew loves art records. The three students who opened this chapter each have a different sweet spot on the musical dimension of novelty, owing to each student's individual makeup of experience, biology, and happenstance. The psychic rewards and penalties that each of these talented youngsters experience from music are dependent on how they view new challenges. Andrew developed a greater appetite for musical risk-taking than did Sheryl, while Omar grew to favor something in between.

Exploring new ideas, new objects, and new situations requires our brains to learn and recognize new patterns, a time- and energy-consuming neural process that demands conscious effort. As with any new experience, learning new musical styles involves a willingness to pay the cost of delayed gratification. The more groundbreaking a record, the harder it is to enjoy on first listen. But when a complex new construct has been encountered several times, it's easier to make predictions the next time you encounter it—and easier to be pleased by it.

We derive pleasure from the way music matches and violates our expectancies. We experience the tiniest quiver of anticipation when we think we know what will happen (when the next drum beat will land, what the next lyrical verse will be about, or where the next melodic line will travel) and satisfaction when our prediction proves correct. When music matches our expectations—when the chorus arrives *just* as the tension peaks—it excites neural circuitry that doles out dopamine, delivering a chemical reward.

But what if the music does not match what we anticipate? Instead of hearing a glorious chorus, what if the drums vanish and all we hear is a vocal or maybe a sustained chord? Or if the last word of the chorus refuses to change or stop, repeating over and over like the word "wrong" on "What We Do (feat. JAY-Z & Beanie Sigel)" by Freeway? Technical innovation like this from producer Just Blaze in 2002 led hip-hop and popular music into the twenty-first century. When we encounter a surprise, something special happens in our brain. The prediction "error" can trigger an even greater release of psychic rewards.

Can trigger: the qualification is important. Not all musical surprises are equally satisfying. As the renowned British biologist Peter Medawar said, "The human mind treats a new idea the same way the body treats a strange protein; it rejects it." The potential reward we experience depends on the exact nature of the surprise: Does an unexpected chord take us to new and fulfilling destinations? Or does it sound bizarre and unwelcome? Does an unanticipated break-down build suspense for an even bigger explosion of music, or does it kill off the groove just as you were getting into it? Some twists work. Some don't. Your satisfaction from a musical surprise depends not only on your familiarity with the genre and your own appetite for having the musical rug pulled out from under you, but whether the surprise satisfies. Listeners like Sheryl are wired to experience fulfillment from subtle or relatively few wrinkles in the classic form of the reggae she loves, while Andrew yearns for the unanticipated whenever he listens.

A clever brain-imaging study from Robert Zatorre's team at McGill University explored the link between musical novelty and mental rewards. While lying in an fMRI brain scanner, each participant listened to a selection of unfamiliar tunes in a variety of popular styles. The researchers observed activity in the listeners' nucleus accumbens, a brain structure involved in generating the experience of reward. After completing their brain scan, each participant was presented with a list of the songs they had just heard. They were asked how much money they would pay to hear each song again, ranging from zero dollars for "I never want to hear it again" to two dollars for "I liked it so much that I'd spend the maximum amount to hear it again." The scientists realized that they could predict how much money a listener would pay to hear a song again, based upon the response of the listener's nucleus accumbens. The greater the activity in the nucleus accumbens as a participant heard an unfamiliar song for the first time, the more they would spend to hear that song again.

Brain-imaging studies like this one illustrate the practical difference between *liking* and *wanting*. *Liking* results from the brain's positive evaluation of a stimulus. *Wanting* is more compelling. When we want something, we spend effort and resources attempting to obtain the rewards it offers. The researchers in the fMRI study showed that during the initial listening period, the auditory brain region and the decision-making brain region both displayed increased activity, exactly as expected when listeners are concentrating on new music. But the intensity of the activity in these "analytical" regions did *not* predict how much money the subjects were willing to spend to hear the songs again. In other words, the perceived novelty of the music—as measured by analytical brain activity—did not predict the intensity of listening enjoyment across all listeners. Instead, each listener exhibited their own personal relationship with novelty, with some listeners feeling that a given level of novelty was "too much" or "too little," while other listeners viewed that same level of novelty as "just right."

The results strongly suggest that there is a "Goldilocks zone" of novelty for every listener, and that this sweet spot can be measured physiologically.

7

If an appetite for risk-taking is going to form in someone's brain, it usually happens in adolescence. Adults with a fully mature frontal cortex (the most important part of the brain involved in decision-making) are typically inclined to slow down and assess the risk involved before trying something new. As we age and take on more work and responsibilities, we have less time and attention available for exploring new things. In terms of mental resource allocation, it makes sense that adults enjoy listening to familiar music, because it frees up cognitive space for more demanding pursuits. Many young people who prefer familiar music do so for the same reason: they direct their adventure-seeking brains elsewhere.

As with your preference for realistic or abstract artworks, whether you are a risk-seeker or a risk-shirker is *not* a reflection of some deep-seated general quality of your personality. Your appetite for novelty, like every other personality trait, is highly contextual. We are all conservative in some situations and adventurous in others. Apple Inc. founder Steve Jobs was one of America's greatest technological visionaries, known for pushing the boundaries of business innovation and creative exploration. Despite indulging in risky gambles in business, when it came to fashion, he famously stuck to a conservative regimen of a black turtleneck and jeans. Celebrity Caitlyn Jenner took an enormous social risk and underwent a very public gender transition, yet she holds many traditionally conservative political views. I enjoy the thrill of the unusual in both film and music, yet I have very conventional tastes when it comes to food.

What influences our individual appetite for risk in music? Thrill-seeking behavior tends to decline with age. For many listeners, the exciting new music we discovered when we were young becomes the reliable playlist we stick with in middle age. There are also a few well-documented genetic factors that influence our individual desire for "sensation-seeking," which psychologist Marvin Zuckerman defines as "the seeking of varied, novel, complex, and intense sensations and experiences, and the willingness to take physical, social, legal, and financial risks for the sake of such an experience."

People who score highly on the Sensation Seeking Scale (a widely adopted measure of individuals' attitudes toward arousal and boredom) tend to *undervalue* risk and anticipate feeling less anxiety over risky behaviors than do the low sensation-seekers. Diving into the mosh pit at a hardcore concert, sharing a boundary-pushing band with your co-workers, and seeking out music that was banned or denounced all involve sensation-seeking with the aim of finding musical love. Though for some of us (like my reggae-loving student Sheryl), the drive to *avoid* boundary-shattering music is what shapes our tastes the most.

Behavioral psychology asserts that if you are rewarded for experimenting with unconventional stimuli—especially when you are young—you are apt to try unconventional stimuli again. In contrast, if your early exposure to unfamiliar experiences resulted in disappointment, confusion, or repulsion, then you are more likely to play it safe the next time you are presented with a similar choice.

Growing up in Southern California, I had access to many radio stations playing a broad variety of music, making it easy to find records on the right side of the novelty-popularity curve. Listening to novel musical forms produced feelings of profound joy in me and forged happy memories. Consequently, I became eager for the next pioneering musical expedition. As I continued through childhood and adolescence, I was rewarded not only emotionally and intellectually for exploring new forms of music but socially as well. In my early twenties, I went to a King Crimson concert with friends. Although the band's elaborate prog rock was not "my music," afterward my reward system said to my risk-aversion system, *I told you so! This was fun!* The feel-good neurotransmitters released from being with excited friends while listening to unfamiliar music reinforced my growing tendency to be open-minded to different musical styles.

On the other hand, I will never forget the first time I tried Vegemite, a malty food spread made from discarded yeast. It was the equivalent of tripping over a rock and doing a face plant. Stunningly unpleasant. A few other youthful excursions into culinary novelty occurred in less-than-ideal social circumstances and resulted in similar reactions. Now it was my risk-aversion system's chance to say, *I warned you! Don't ever do that again!* Perhaps it was simply the luck of the draw, but I was learning through experience to play it safe when it came to food, even while my musical risk-taking was reaping rewards. Maybe if my early food experiences had exposed me to exotic dishes that I had naturally enjoyed, or if I had tried unfamiliar cuisine in settings where I received warm social reinforcement, an appetite for epicurean adventure might have taken hold. But that's not what happened. As a result, I quickly came to embrace my own limited set of comfortingly familiar dishes. Today, I go with "the usual" nearly every time I dine out and "the latest" when searching for a new musical artist to enjoy.

Of course, when it comes to the music you love, it doesn't matter *how* you ended up liking what you like. Whether you're on the far right of the curve, the extreme left, or stationed firmly in the middle, it indicates absolutely nothing about the depth of your passion for music. It merely provides another window into the music of you.

Tone Deafness

Tone deafness is an impaired ability to discriminate between different musical pitches. A tone-deaf person often fails to recognize familiar melodies or detect when a familiar song has an off-key note. Tone deafness is much more common than absolute pitch, manifesting in just under 2 percent of the population.

People with tone deafness have impaired singing abilities. Although they may be able to reproduce a note they've just heard, they frequently report that they are unsure if it is on pitch or not.

Technically known as *amusia*, tone deafness does not usually impair speech perception, although there are exceptions. About 30 percent of tone-deaf listeners have difficulty distinguishing an assertion from a question because they cannot determine whether the pitch of the final syllable is rising or falling.

Though true tone deafness is fairly rare, many people who are poor singers incorrectly believe they are tone-deaf when in reality their deficiency is in *producing* tones correctly, rather than *perceiving* tones correctly.

CHAPTER 4

MELODY

This Is What Music Feels Like

✧

What do you suppose they mean when
they say "it's not melodic?"
Isn't any string of notes a melody?

—*Leonard Bernstein*

1

"THE 500 GREATEST ALBUMS OF ALL TIME" WAS A LIST published by *Rolling Stone* magazine in late 2020. The year-long project gathered nominations and votes from artists, songwriters, producers, music critics, and industry executives, with the tacit agreement that public opinion—meaning record sales—counted, too. Scrolling through the final list provokes some *What?!* reactions (*400 Degreez* by Juvenile, edging out both *The Stooges* and Bonnie Raitt's *Nick of Time*) and *I never woulda guessed!* (Alice Coltrane's modal jazz album *Journey in Satchidananda* placing higher than *Coal Miner's Daughter* by Loretta Lynn) and *YES!* (Nirvana's *Nevermind* at No. 6). What is indisputable is that every one of the five hundred albums on the roster earned the devotion of a wide swath of listeners. What was it about the patterns encoded on these albums that caused so many different brains to fall in love with them? Is there any way to classify what, exactly, listeners' minds were focused on as they fell for these celebrated works of musical art?

So far, we've explored three dimensions of our listener profile that are *not* specific to music: authenticity, realism, and novelty. Each of these dimensions is processed by multiple brain networks simultaneously, rather than activating a single modality-specific network. Brain networks that process each of these three dimensions influence not only our emotional response to records but our response to all forms of creative art, including movies, novels, and dance. Thus, we might call authenticity, realism, and novelty the *aesthetic dimensions* of our listener profile.

In the next four chapters, we're going to turn our attention to four dimensions of our listener profile that *are* specific to music: *melody*, *lyrics*, *rhythm*, and the underappreciated dimension of *timbre*. There are two important differences between our aesthetic dimensions and the four *musical dimensions*. Each of the aesthetic dimensions is binary. Sweet spots on the aesthetic dimensions can be imagined lying on a single axis running between two opposite poles: above-the-neck versus below-the-neck, realism versus abstraction, novelty versus familiarity.

Musical dimensions, in contrast, are not binary. Our perception of melody includes several distinct melodic features, each with its own axis. (Technically, the four musical dimensions of our listener profile are actually "musical *spaces*," but for clarity and consistency, I'll continue to refer to them as dimensions.) Melody, for example, can have a wide note range or a narrow, monotonic range. Melody can be expressed in a style that is staccato (brisk and precise, with each note sharply detached from the previous note) or legato (smooth and flowing, with notes connected to each other and sometimes overlapping). Melody can evoke specific emotions by mimicking speech, or it can sound obscure and therefore open to broad interpretation. Thus, you likely possess multiple sweet spots on your melody dimension, each spot corresponding to a different melodic feature. In the same fashion, you may also possess multiple sweet spots on the lyrics, rhythm, and timbre dimensions of your listener profile.

Another key difference between the aesthetic and the musical dimensions is that the musical dimensions, unlike the aesthetic dimensions, are each processed by a single specialized brain network. Each network generates a distinct mental reward. We might say that melody serves as a record's *heart* for how efficiently it ignites your emotions. Even simple melodies can evoke nuanced feelings like wistfulness, pride, adventure, or unrequited love. In contrast, the words of a song tap into your brain's knowledge system, so lyrics serve as a record's *head*. Rhythm becomes a record's *hips*. The groove of a record engages your brain's motor system and compels you to move. Timbre is the raw quality of a musical sound that constitutes its identity, such as the sharp buzz of a saxophone, the resonant hum of an acoustic guitar, or the drone of a didgeridoo. That's why timbre serves as a record's *face*.

When songwriters or producers evaluate a new song, they can use a simple exercise to decide which of these musical dimensions might be harnessed to deliver the biggest reward to listeners. *Is it easy to hum in the shower?* If so, this song has a good melody. *Does it work on the page?* If so, this song has strong lyrics. *Does it pop into your head while you're exercising?* If so, the song's best feature may be its groove.

When you listen to a record, each musical dimension has a shot at earning your affection, especially if a record's best feature aligns with your own favored dimensions. Music lovers whose passions run hot for great lyric writing may have Leonard Cohen, Patti Smith, Nas, Alex Turner, or Hank Williams records in their music libraries, because these lyricists have reached a high bar of excellence. Those preferring records that groove the hardest may have a decent-sized African or Latin music collection. When multiple elements on a record match up with multiple sweet spots on your listener profile, that record can make you fall head over heels in love. That's precisely what many listeners experienced while listening to the *Rolling Stone* poll's top vote getter: Marvin Gaye's masterpiece *What's Going On*.

His 1971 album is a pained call to halt the social discord that was pulling America apart near the end of the Vietnam era. For many listeners, the title track, "What's Going On," delivers high-octane rewards on three of the primary musical dimensions: the song features a heart-wrenching melody, socially poignant lyrics, and an irresistible rhythm that interplay in perfect musical compatibility, thanks to Gaye's artistry.

The drums and congas are upbeat, jaunty almost. The rhythm guitar joins in with the chord progression—adding tension and anchoring the rhythm section. A saxophone calls out the melodic theme, then steps back and lets Marvin take over in a voice as smooth as the softest suede, singing the opening line: "Mother, mother, there's too many of you crying." Then the bass does what bass does best: treads the boards between a rhythmic role and a harmonic role, adding the subtle shade of secondary emotions—worry and hope—in equal measure. Behind it all the strings raise a dominant melodic cry, wordlessly begging us to stop fighting each other. Marvin tells us that "we've got to find a way to bring some understanding here today." The gentle backing vocals let Marvin take the spotlight, but as the record continues, they split into factions, chattering away and abandoning the script, just as a chaotic, diverse society does. Then, redemption!

All the voices come together as the song approaches its end. The strings glide faster, like birds, going higher and higher yet never getting away from the tonal center, a relentless pressure scaling its way upward, toward release. For many listeners, our brain networks handling melody, lyrics, and rhythm all respond in the sympathetic resonance of connection. Your body can sink into the groove and lock onto its inviting rhythms, your heart can get tugged into the deep end of the emotional waters by the melodic strings, and your mind can sadly recognize that fifty years after the record's release, its lyrical message is still woefully timely. (We're ignoring timbre because Marvin Gaye made his records prior to the era of timbral innovation ushered in by DAWs.)

Over the next four chapters, we will explore how every one of the four musical dimensions of your listener profile holds the potential to bewitch you—and help you pinpoint where your own sweet spots lie.

2

In 1940, a twenty-four-year-old Frank Sinatra desperately wanted to be the greatest singer in the world. He was trying to achieve this goal by emulating the man who *was* the most popular singer in the world: Bing Crosby.

Crosby was the first major singer to adjust his vocal style to suit the new sound-amplifying technologies that were taking over concert halls and radio. Previously, singers needed lung power and a "false vocal fold" —a partial constriction in the larynx just above the actual vocal folds—to project their voices over the band and reach listeners at the back of the hall. But with the development of electronic technologies, singers could lower their voices all the way down to a whisper, knowing that the microphone would pick up even their softest exhalations. Crosby was the first of these "crooners." He exploited the new technology to forge a laid-back, "cool" vocal style that thrilled women and made men want to be him, including an ambitious young Sinatra.

By 1940, Sinatra had already achieved moderate success. Though he was not yet an A-list celebrity, his songs were played on the radio and he had a steady gig as the lead vocalist for the Harry James Orchestra. This was the big band era, when the star of the show was the bandleader—not the singer. The bandleading Count Basies, Duke Ellingtons, and Tommy Dorseys of the 1940s were comparable in popularity to the Mick Jaggers, Janis Joplins, and Sam Cookes of the '60s: virtuoso musicians who were household names. In contrast, a big band singer like Sinatra was perceived as just another instrument in the troupe. Sinatra yearned for greater recognition, however. In his early years he pursued this objective by aping Crosby's mellow approach to crooning.

You can hear the smooth, sentimental, yet clearly derivative version of Sinatra on the 1940 recording of "All or Nothing at All" with the Harry James Orchestra. Sinatra had a very good voice, but there was nothing particularly special about his singing style. Three decades later, Sinatra's vocals had changed drastically. Listen to his 1971 retirement concert performance of the same song, "All or Nothing at All," at the Ahmanson Theatre in Los Angeles. You don't have to be a music expert to feel how Sinatra's later version is more engaging, authentic, and unique. What changed? Long before 1971, Frank Sinatra had become a master of *melody*.

Sinatra's ascent to melodic genius was launched in 1940 by a visit to Carnegie Hall, in New York City. He was always listening for new ideas that might improve his singing. On that night, he attended a classical concert of Brahms, Debussy, Rachmaninoff, and Ravel performed by the legendary violinist Jascha Heifetz. Sinatra was enthralled. Heifetz's bowing technique produced astonishingly long and evocative melodic phrases that led the crooner to a musical epiphany.

Sinatra observed that Heifetz would "get to the end of the bow and continue without a perceptible missing beat in the motion." Heifetz could make a note go on and on and on without interruption by keeping the bow moving over the vibrating strings in a steady circular motion. Sinatra wondered if the same technique could work with the human voice.

The young Sinatra took up swimming, running, and listening to classical music, all the while practicing vocal exercises that enabled him to increase the endurance and energy of his singing. He sought assistance from vocal coach John Quinlan, who helped him duplicate with his vocal apparatus the melodic feats Heifetz accomplished with his bow.

Breath control and the ability to produce longer vocal phrases gave Sinatra more ways to adjust his timing and enunciation. Over the course of his decades-long career, Sinatra finessed and mastered the timing of melodic phrases to "make good and goddamn sure" that the audience paid attention to every single word he sang. He infused each word with intense feeling. He precisely controlled when he started and finished each phoneme (for example, stretching out the word "all" in "all or nothing at all" to distinctly emphasize both the "a" and the "ll" sounds), choosing to place some notes ahead of or behind the rhythm section to set the emotional pace. This ensured that the band followed Sinatra's lead, and not the other way around.

James Kaplan recounts in his consummate two-volume biography of Sinatra that audiences, musicians, and bandleaders all agreed: the skinny kid from Hoboken had developed a unique vocal character. Sammy Cahn, the legendary songwriter who penned some of Sinatra's biggest hits, said of his singing prowess, "Frank can hold a tremendous phrase, until it takes him into a sort of paroxysm—he gasps, his whole person seems to explode, to release itself."

You can hear Sinatra's mastery of melody on the 1966 live recording of "It Was a Very Good Year," on the album *In Concert: Sinatra at the Sands*. Sinatra creates phrases that make us hang on every single inflection in the song's story. Pay close attention to the rhythm of the Count Basie Orchestra under the direction of young Quincy Jones. Listen for where Sinatra places his own phrases in relation to the band's phrases. He is ahead of the downbeat when he wants to emphasize his fervor about life's romantic stages from young to old, but he drags behind the beat when he wants to let the superb musicians take the lead. In particular, pay attention to his timing when he describes being seventeen and then listen to how his timing changes when he sings of being in the autumn of his years. As the story progresses, none of the ardor goes away, just the hurry. He controls his pitch to hug the curves of the melody like a race-car driver, expertly skidding around "a veeeerrrrry good year" after the line "Then I was thirty-five . . ."

"Declination" is a term of art that refers to a drop in pitch signaling the end of a melodic phrase, just like the drop in pitch that signals the end of a spoken sentence. Consistent declinations are the mark of a novice singer. Beginners tend to expel most of their vocal energy at the start of each melodic phrase and run out of power before they reach the end. Not Sinatra.

Under the tutelage of Quinlan, Sinatra learned to control melody and release any word with the exact amount of power he chose for it. Sinatra could make phrases last longer than the listener expected (delivering a subtext of virility) and would sometimes run one phrase smoothly into the next phrase without seeming to pause for breath, just like Heifetz's bowing (a subtext of even *more* virility!).

Sinatra's melodic technique was effective at commanding his listeners' attention, empowering him to manipulate the emotions they experienced while hearing him sing.

3

In simplest terms, a melody is a sequence of musical pitches. There is something special about the way our brains treat sequences of pitches, compared to sequences of anything else. Read this sequence of words: green-blue-pink-green-blue-pink-orange-purple-purple-green-orange-purple-purple-green, while imagining each color as a flash of light. Did you feel any emotions as you looked over the sequence? Most people do not. Can you reproduce the sequence after a single viewing? Again, probably not. Yet the same sequence expressed as sound—where each color matches up with a distinct musical pitch—serves as the melody for "Three Blind Mice," a frisky children's tune that most people can repeat back after a single listen.

The up-and-down shape formed by a sequence of musical pitches is called a *melodic contour*. A contour might be: *three notes up, four notes down, two notes up, three notes down*. Melodic contours are less precise than *melodic intervals*. Two different melodies can have the same melodic contour because their interval sizes—the exact spacing between the notes in each melody—are not specified. A contour is a kind of "rough sketch" that our brain easily encodes when first memorizing a melody.

As illustrated in the following figure, if we draw the melodic contour for a particular tune, we can get a general sense of the melody. The opening bars of "Over the Rainbow" feature a melodic contour

Melodic Contours

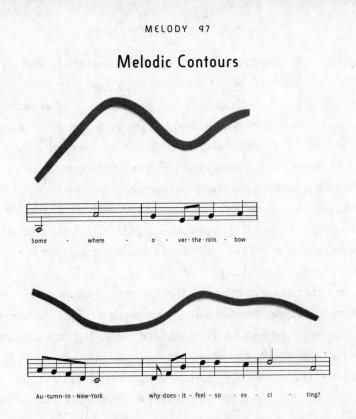

with a dramatic ascent, jumping up a full octave before falling and getting relatively narrow. The opening bars of "Autumn in New York" feature a gentle descent, followed by smoothly ascending notes. The shape of a song's melodic contour provides some hints to the kind of emotions the melody might evoke in a listener. An ascending melody can give a sense of rising excitement. This helps "Over the Rainbow" set a dramatic tone. A descending melody can feel poignant or nostalgic. This is beautifully expressed by "Autumn in New York."

The melody on a record is usually performed by the lead instrumentalist or lead singer. Pianist and vocalist Nat King Cole is considered an undisputed great of the jazz era, and one reason was his deftness with melody. Listen to how he treats the vocal lines in "Nature Boy" in his original 1947 recording. The piano intro is unremarkable, keeping the stage dimly lit for the singer's entrance. The melody itself is simple, sweet, and poignant. Like "Over the Rainbow," it starts with an ascending pair of notes, but this melody tends to repeat short phrases, heard on the lyrics "very far, very far, over land . . ." As with Sinatra, we hang on to every one of Cole's vocal phrases. Cole caresses the melody to express emotional subtlety. Hear how the piano and strings pick up the baton when they take over the melody in the solo section after the second verse. The simple and undistracting piano solo—played by Cole—reiterates the main melodic theme to help listeners focus on it rather than the lyrics or rhythm in this song.

Another sequence of pitches that frequently appears in music is *harmony*. A harmony is like a shadow to the main melody. Arrangers sometimes compose a harmony line that follows the same contour as the main melody, enriching and augmenting it. Jazz musicians are experts at harmonic improvisation: "composition in real time." Typically, a jazz musician starts a solo by playing just enough of the song's melody that we can recognize it before veering off to play his own improvised harmonic notes over the same chord progression. You can hear a sterling example of harmonic expertise on another version of "Nature Boy," this one recorded in 1965 by the great American saxophonist John Coltrane. (The recording was posthumously released in 2018 on *Both Directions at Once: The Lost Album*.)

When Coltrane's sax enters, it plays just enough of the melody to give us a frame of reference. Most listeners in the 1960s were familiar with "Nature Boy" because it had already been a hit for several artists, including Nat King Cole, so Coltrane's aim here is to get us to *reconsider* the well-known tune. He does this by playing unexpected harmony notes that seem to tell an alternate version of the tale. Coltrane's harmony suggests that the "nature boy" was perhaps not quite as sweet as the beloved melody would have us believe, hinting that he came to be a "strange, enchanted boy" through a darker journey. If the melody supports the lyric "The greatest thing you'll ever learn / Is just to love and be loved in return," then Coltrane's inventive harmony counters with a suggestive backstory for the boy who "wandered very far."

Sometimes harmony features a different contour than the main melody, going off in a new direction to provide subtext or counterpoint. This approach lets a record maker express more than one emotional tone, like casting two dissimilar actors in a "buddy picture." If you listen to James Taylor's 1970 classic "Fire and Rain," you can sing along with the bowed double bass (the low strings) in the chorus to hear how its harmony notes both complement and contradict the vocal melody. Listen to how odd and intense the bass becomes as it reaches the top of the third verse (around the two-minute mark, with "Been walkin' my mind to an easy time"). The dark and insistent notes add a somber subtext to the song—but why? The melody expresses a soft feeling of peace and guarded hope that James might see Suzanne again, but the bass-driven harmony intimates that something dire has happened. The tension between melody and harmony endows this record with a depth of feeling, like a stage play that mixes comedy with tragedy.

Contrasting the melody with a divergent harmony, as in Coltrane's "Nature Boy" and Taylor's "Fire and Rain," can be a fine way to add nuance to a record, but immense emotional power can also be achieved when a record's melody, harmony, and chord progression—the tonal skeleton undergirding the melody—all team up to say the same thing. A good example is the Allman Brothers classic "Midnight Rider." There was a time when leader Duane Allman was considered the American South's most expressive and soulful guitarist. His brother, Gregg, wrote this three-minute song about running from the law, drawing on American folk, blues, R&B, and country music. Listen to how the melody functions as a declaration of who the singer is: an outlaw. The chord progression is insistent. After Gregg sings the verse telling us of the state he is in, the chords never change. They remain steadfast, echoing the proud and resolute melody. (This is how a lot of R&B music works, by staying in one lane and letting feeling build up through persistent repetition.)

Another prodigy of the American South is producer and songwriter Pharrell Williams, whose brilliant single "Happy" was a well-deserved smash hit in 2013. If you're not familiar with the song, listen to it all the way through to the end of the chorus, which starts: "Because I'm happy . . ." Now sing the melody of the chorus but sing "la la la la" in place of the actual lyrics. You may be surprised. On its own, the melody doesn't carry the entire carload of joy. Rather, it is the drums, bass, lyrics, melody, and harmony working in tandem that deliver the irrepressible feeling of utter elation. It is the *record*, rather than the *song*, that expresses happiness through the accompanying chord progressions (as heard in the exuberant guitar framing the verses) and the sky-high harmonies of the backup singers.

Once you've tried out this exercise with "Happy," try the same thing with the chorus of Carly Rae Jepsen's infectious 2012 hit "Call Me Maybe." Sing "la" in place of the lyrics, "I just met you, and this is crazy / But here's my number, so call me, maybe." This time, you may agree that the melody alone conveys the cheerful feeling in the words: happiness with a touch of hesitancy at the idea of taking a risky step.

For all its simplicity, melody can rouse a sweeping range of emotions. Somehow, a sequence of tones rising and falling evokes feelings of rich sorrow, quiet pride, or surging triumph. What is going on in our brains when a melody causes us to *feel*?

4

In the early 1990s, I conducted an armchair experiment with my new Boston terrier pup, Gina. Before setting down her food bowl, I would sing the melody "O Christmas Tree" with "di-di-di" in place of the words. When it was bedtime, I sang her "I'm Late" from *Alice* in *Wonderland*, again just singing "di." Before heading out to visit my family, I sang a tune from an old Betty Boop cartoon about visiting Grampy. Our walk to the dog park was prefaced with the melody "Copacabana."

I wanted to see if she could learn melodies. Would Gina head in the right direction (kitchen, upstairs, back door, front door) when she heard the relevant tune? I figured it would take months of training. In fact, it took about a week. All I had to do was sing the familiar melody and her eyes would grow big. She'd dash back and forth between me and the correct location with the happy face of anticipation that every dog owner knows. You might think that her behavior was due to the timing of the activities rather than the melodies, but I can say with confidence that it wasn't. I was a very busy record maker, working long and erratic hours without a steady routine. Suppertime, bedtime, and outdoor fun was catch-as-catch-can. Gina heard a melody and knew what activity it signified.

Like Gina, many species of animals possess the ability to recognize meaning in a sequence of tones. Yet animal-communication experts describe an odd asymmetry in how other species produce, use, and comprehend vocalizations. The renowned ethologists Robert Seyfarth and Dorothy Cheney write that "learned, flexible vocal production" is found in only a few orders of birds and some mammals. The ability to modify an innate set of vocalizations over an animal's lifetime is rarely seen. Unlike humans, who voice an ever-changing variety of verbal expressions, the barks, hoots, whinnies, roars, and bugles in the animal kingdom tend to be consistent from youth to old age.

But *comprehension* of vocalizations, even in non-human animals, is quite flexible and can be modified by life experience. Many species possess the neural infrastructure to learn the meaning communicated by a sequence of tones. Gina could not *produce* a melody the way I could, but she had no trouble distinguishing between different melodies, and she could execute different behaviors in response to each one. In early humans, the ability to associate melodies with intentions and feelings got finely honed.

An ongoing question in evolutionary neuroscience is: Which came first, music or language? The available evidence seems to point toward their coevolution, the development of one supporting the development of the other. Many species are known to express and elicit emotions through their melodic calls, leading theorists to coin the term "musilanguage" for early human vocalizations that communicated both information and feelings. This idea makes intuitive sense. Neuroimaging studies have shown that our brain's activity when processing social signals (such as body language or the tone of someone's voice) is very similar to its activity when processing musical signals. In humans and other animals, producing a sequence of high-pitched, clear tones is common in happy or prosocial contexts, while producing low-pitched, noisy tones are more common in aggressive or threatening contexts. Short calls with ascending pitches have an arousing effect, while long calls with descending pitches are calming. Dissonant melodic intervals, such as the two alternating tones produced by many emergency-response sirens around the world, are fear-inducing. We instinctively draw upon this knowledge when we use vocalizations to communicate with working animals such as dogs and horses.

When scientists study animals' responses to melodies, they usually play human music to their subjects. But what if scientists instead played "animal music"? What if researchers composed music that featured a species' own emotionally relevant tempos, timbres, and melodic intervals? Would a monkey, for instance, find monkey music more affecting than human music?

Researchers Charles Snowdon and David Teie explored this question with cotton-top tamarins, a species of monkey named for the distinctive poof of white hair adorning its head. Cotton-top tamarins live in Colombia, South America. They're about the size of a squirrel. They prefer silence to human music, although if forced to listen, tamarins will opt to hang out in an enclosure that offers a Russian folk lullaby played on a flute over one that broadcasts the highly arousing "Nobody Gets Out Alive!" by the electronic musician Alec Empire. Tamarins' lack of interest in music makes them the perfect species for testing the hypothesis that an animal's emotional response to melodies might be stronger and more salient when they listen to melodies that match the species' natural auditory template.

Snowdon and Teie created two sets of melodies on a cello, enhancing certain harmonic frequencies to put them within the tamarins' range of hearing and vocal timbre. Their buoyant, encouraging compositions featured melodic intervals similar to those made by tamarins in positive social situations. Their fearful, threatening compositions featured faster tempos and melodies that ascended in pitch, similar to how tamarins vocalize when frightened. To the researchers' delight, the monkeys responded to these customized melodies as they'd hoped: by exhibiting increased arousal and alertness when they heard the threatening melodies and exhibiting calmer, prosocial behavior when they heard the melodies that sounded like a peaceful day in the jungle. The same team performed a similar study with cats and showed that felines also prefer music that approximates the melodic contours of their own vocalizations.

There's an important difference in how humans and non-humans use vocal melodies, one that underscores our unique musical intelligence. A human can use her voice to induce an emotion in another human, *even if she is not feeling the emotion she wants to elicit.* A frazzled, anxious mother can sing a soft, gentle melody and calm her infant. An exultant award-winner can choose to appear humble at the podium, consciously deploying vocal intonations that convey humility. Speech prosody—the melodic and rhythmic stresses we put into our voices—conveys emotions (sometimes spontaneously, sometimes guardedly, sometimes duplicitously), letting others know, "I feel you."

Regardless of when our human auditory circuitry evolved, the audio signals that impinge on our eardrums are routed to several parallel networks in our brain, each focused on a distinct quality of the soundwave, including its emotive tone. One of these networks processes the acoustic pattern of melody in the soundwave. Another network processes the acoustic pattern of words.

This accounts for the middle-aged suburban dad sitting at a traffic light while belting out with Aretha Franklin, "YOU. MAKE. ME. FEEL. LIKE. A. NAT-UR-AL WO-MAN!" The odds are high that Dad isn't feeling the *lyrics* of female empowerment so much as he's feeling the confident *melody*. The simultaneous dual processing of melody and words allows Dad's mind to focus on either the intonation or the information, and it chose the intonation. (Our brain's automatic division of musical inputs lends credence to many melody lovers' claim that they "never listen to the words.")

One of the first scientists to demonstrate this intriguing neural split between melody and speech was the music psychologist Diana Deutsch. One day as she sat at her desk dictating research notes into her tape recorder, she spoke the sentence "The sounds as they appear to you are not only different from those that are really present, but they sometimes behave so strangely as to seem quite impossible." When she played back the recording, she noticed something unexpected. Curious, she set about conducting an experiment to examine the oddity more carefully.

Deutsch presented her recorded sentence to two groups of American college students. Group A heard the phrase "sometimes behave so strangely," then obeyed her instruction to mimic exactly what they heard. They repeated the phrase back accurately, albeit with droll attempts to copy Deutsch's British accent. Group B also heard the phrase "sometimes behave so strangely"—but with a twist: for these listeners, the phrase was repeated several times on a continuous loop. When the Group B listeners were asked to repeat back what they heard, they *sang* the phrase. The exact same phrase was heard as *speech* by some listeners and *melody* by others.

You can try the experiment for yourself at our website, ThisIs WhatItSoundsLike.com.

What was going on? According to Boston University neuroscientist Stephen Grossberg, our brains separate, or "factorize," the acoustic stream of Deutsch's voice into two distinct neural pathways, one processing melody, the other processing speech. Initially, our minds prioritize the informational content of the speech—we focus on the *meaning* of Deutsch's words. This is not surprising: speech is, by far, the most important sound we humans hear. But our speech-processing pathway operates with a useful quirk.

To ensure that our minds can flexibly deal with ambiguities in speech, the speech pathway automatically "forgets" the initial meaning of words through a process known as habituation. (The same process of habituation also makes us gradually stop noticing the smell of squeezed lemons as we cook or the patter of rain on the roof.) Speech habituation accounts for the "Wait, what?!" reaction by helping us reinterpret words when our initial interpretation may have been wrong, such as taking a moment to realize that the newspaper headline "Fireman Helps Dog Bite Victim" doesn't mean "the fireman enabled the dog to bite the victim" but, rather, "the fireman aided the man bitten by the dog." However, an unexpected side effect of speech habituation occurs when an identical phrase is repeated over and over.

Because of habituation, our minds gradually forget the initial interpretation of the spoken phrase "sometimes behaves so strangely." Instead of reinterpreting the repeated phrase with a new linguistic meaning (after all, we successfully understood it the first time), our mind gradually stops focusing on the phrase's meaning and shifts its attention to another salient dimension of the acoustic pattern—its melody. Melody is processed by a pathway in our brain that is older and more primal than the speech pathway. Once our mind switches focus to the intonation of an iterated snippet of speech, we now hear its melody through all future repetitions, having "forgotten" its linguistic meaning. After you have heard "sometimes behaves so strangely" as music, you will automatically hear it that way, even if you don't encounter the phrase again until months later.

Deutsch's research helped make one thing clear: your brain processes melody independently of how it processes lyrics.

5

Baby Joris, a German newborn, was only four days old when a researcher with a microphone waited patiently for him to let the room know how he was feeling. As Joris's mother changed his diaper, he obliged by offering up a few unhappy wails. Mom smiled at the researcher, Kathleen Wermke, and remarked that her son's wail sounded German. Wermke agreed.

Wermke wasn't simply offering polite encouragement to a proud mother. A few years earlier, her research team had analyzed a half million newborns' cries gathered from around the world. Wermke made the remarkable discovery that shortly after birth, babies appear to cry in their native tongue. French infants tend to cry with a rising melodic contour. German infants tend to cry with a falling contour. These disparate contours match the disparate prosody of French and German speech.

In the final trimester of pregnancy, a fetus's auditory system has developed well enough that some muffled sounds are audible as the baby floats in its gestational soup. Brain development at this stage is just complex enough to allow us to learn simple melodic phrasings, such as intonations, which is one reason why newborns prefer the familiar sound of their mother's voice over the sound of an unfamiliar voice. Once we enter the world of airborne sounds, the speech prosody of our family's native language influences how we hear and, shortly thereafter, how we produce sounds. A newborn's ability to imitate speech, even by waking the neighbors with a long, drawn-out WAAAAAAH!, is fundamentally linked to the human sense of melody.

In the first few months of life, newborns' melodic cries become increasingly intricate and begin to resemble the musical intervals prominent in their cultural environment. Dr. Wermke writes that "in crying, elementary constituents of both musicality and language faculty are unfolding." In fact, there is a link between the melodic intonations babies learn in the womb and the type of music those babies will eventually compose, should they become music makers.

Among Western languages, French has more frequent melodic peaks than British English. Experimental findings have shown what music and linguistic scholars have long suspected: that instrumental music from France *sounds* French, just as music from England *sounds* English. Tonal languages such as Mandarin and Vietnamese (where pitch changes determine the meaning of certain words) tend to have wider intervals and more frequent contour changes than non-tonal languages—and so does the music composed by tonal language speakers. Although many music listeners seek out and enjoy melodies from around the world, the melodies of our native tongue sound the most familiar and speak the most passionately to us, because our minds were soaking up these melodic contours before we were even born.

The fact that we have a deep preference for the melodies of the native culture we are born into means that we may be naturally prone to disliking unfamiliar melodies, a fact supported by the empirical observation that people express less affection for what they consider to be "foreign" music. Sadly, this distinction has enabled the military to wield music as a weapon of coercion. Uncooperative Iraqi prisoners held captive during the Iraq War were persuaded to talk to interrogators after long barrages of heavy metal music, such as Metallica's "Enter Sandman," alternated with American children's songs, including the purple dinosaur Barney's theme, "I Love You."

While retaking the city of Fallujah, American soldiers blasted their favorite rock and rap music from loudspeakers mounted on gun turrets. This tactic was described by one soldier as a sonic "smoke bomb," adding that "our guys have been getting really creative in finding sounds they think would make the enemy upset." They were not the first Americans to use music as a cudgel. The Panamanian dictator Manuel Noriega, an opera lover, surrendered after his Vatican embassy refuge was blasted for a week with high-volume AC/DC, Mötley Crüe, Led Zeppelin, and other rock music.

When it comes to our listener profiles, one listener's green zone can be another listener's circle of hell.

6

Where do your own sweet spots lie on the dimension of melody? To focus attention on the kinds of melodies that are personally rewarding to you, our record pull departs from the listening room and heads to the cinema.

Most cinematic scores feature a vivid and memorable melodic theme written to express the emotional tone of the film. Because movies showcase the entire palette of human emotions, there is great variety in cinematic scores. Movie themes rarely incorporate lyrics, which would distract from important action in the movie, including dialogue between characters. Of course, lyrics improve our memory for melody, and so, to ensure that moviegoers can hum the theme the next day, film composers strive to write themes that can "speak" emphatically without words. As a result, movie themes are a great resource for exploring your melodic sweet spots.

Let's examine three distinct axes that contribute to the melody dimension of your listener profile: melodic range (wide versus narrow), articulation (legato versus staccato), and complexity (simple versus complex). You may (or may not) possess a sweet spot on each of these melodic axes. Let's begin by considering melodic range, including wide pitch intervals, which soar from low notes to high notes and back again, and narrow melodies where only a small set of neighboring pitches is heard.

Personally, I like romantic melodies featuring sweeping intervals, as heard in Michel Legrand's theme from *Summer of '42*. I was fourteen years old in 1971 when my mother passed away after a years-long illness. *Summer of '42* was released that year, and its musical theme seemed to calm my emotional vertigo. The movie takes place on Nantucket during World War II and tells the story of Hermie, a fifteen-year-old boy with a crush on a young bride whose husband is off fighting the war. Although I was coping with something other than lovesickness, the melody suggested a personal pain intermingled with a bigger tragedy of adult loss, shot through with beauty. Part of what makes this composition so poignant is that each melodic phrase seems to end too soon, perfect for a tale about a soldier far away from home who is doubtless not much older than Hermie but forced into experiencing a far, far different life. The juxtaposition of both the natural and the unnatural in these young lives is expressed in the wide-ranging melody that is profoundly sad, yet uplifting.

My coauthor admits to a fondness for narrower melodies, the kind that typically accompany techno or rap music, as heard in Philip Glass's theme from the movie *Koyaanisqatsi*. Its minimalist melodic contour remains within a constricted range of notes, employing the repetition of arpeggios that tonally never travel very far from where they begin. The experimental film, released in 1982, eschews characters and narrative in favor of slow-motion and time-lapse footage of cities and landscapes. Because the movie lacks any dialogue, Glass's insistent, evocative soundtrack plays an outsized role in guiding our emotional reaction to the film. The melody slowly but steadily gains in speed and volume as the movie's images progress from peaceful though awe-commanding natural landscapes to the frenetic hustle and traffic of human-thronged metropolises. By staying within a narrow range of notes that endlessly cycle, Glass's melody almost feels like the increasingly agitated vital signs of a living planet—the breath and heartbeat of an Earth spiraling out of control.

Articulation is another axis of melody where you may have a preference. Perhaps you like the melodic theme from *A Beautiful Mind*, featuring legato notes that flow into one another. The film traces the life of a Nobel laureate mathematician who battled schizophrenia. Its gentle melody suggests fragility but also an enigmatic yearning to grasp something just beyond reach. The main melody seems to ascend without ever getting higher, like an M. C. Escher painting or the acoustic "barber pole illusion," called the Shepard tone. Ethereal female voices loop around flutes and lilting strings while other strings pulse underneath. James Horner's theme evokes the feeling that even something beautiful can hurt if you have too much of a good thing.

I have a fondness for flowing melodies, while Ogi warms to staccato melodies with some separation between the notes, as in the theme to *Cool Hand Luke*. The bittersweet melody is plucked on the strings of a guitar. The movie stars Paul Newman as a prisoner struggling against the confines of a Florida prison camp. He is eventually shot and killed when he tries to escape. The theme's melody echoes the plot of the movie, starting out with a steady, unrushed, and yearning feeling which gradually becomes more and more anxious and disrupted, until the melody ends on a dark, haunting, and ultimately unresolved harmonica note. The movie's score, composed by Lalo Schifrin, was nominated for an Academy Award.

You may possess a sweet spot on the axis of melodic complexity. Some listeners relish melodies that dive through various moods, such as Danny Elfman's theme for the Tim Burton movie *Edward Scissorhands*. The music begins tenderly, proceeds through some adventurous melodic arcs, slows down to express longing, and ends on a feeling of naïveté imbued with tension, as if someone's innocence may be about to come to an end. And, indeed, the movie tells the story of the enchanted Edward Scissorhands, a human-like boy with scissors for hands, magically created by an old inventor. Edward descends from the inventor's mansion on a hill into a California suburb where he makes friends and enemies, falls in love, and is eventually chased back to the mansion, where he might be doomed to live alone forever. Elfman manages to capture the dramatic twists and turns of the plot with an intricate melodic theme.

Other listeners are fond of melodies that never stray far from a simple compelling phrase, perfectly typified by Maurice Ravel's "Boléro," used in a romantic scene from the 1979 movie 10. Legend has it that Ravel played the melody on piano for his friend, using only one finger. He suggested that something about the phrase begged to be repeated. His composition does just that, repeating the melodic theme over and over again, building up the instrumentation on each iteration, as the insistent rhythm grows louder and more dramatic. The melody never changes, but the composition builds in potency through layered orchestration and dynamic percussion.

Think about the cinematic themes that have lingered in your head the longest after the movie has ended. Look for consistencies across these melodies, for they may reveal your own sweet spots. If you have a particular fondness for movie themes, this may be a good indication that melody is a vital dimension of your listener profile.

Synesthesia

Synesthesia is a neural condition where one type of perception, such as sound, simultaneously produces the experience of a different type of perception, such as vision. Tone-color synesthetes tend to "see" a specific color when they hear a specific pitch. Less commonly, synesthesia can involve other forms of cross-modal perception, such as smell, taste, and touch.

The exact neural mechanism supporting synesthesia remains controversial, but evidence points to an origin in early childhood experiences. A young child with a colorful toy xylophone or an interactive music instruction primer may learn to associate the colors of the keys or the lesson pages with the sound of the notes. If the association is strong enough, neural connections form that link the auditory and visual networks to each other and to the networks involved in verbal categorization. These neural connections generate the experience of synesthesia.

CHAPTER 5

LYRICS

This Is What Identity Sounds Like

✧

Sometimes songs are not what they were meant
to mean, but rather what they need to
mean to someone.

—*Bono,*
"60 Songs That Saved My Life"

1

THE BRITISH AUTHOR A. A. GILL, KNOWN FOR HIS audacious wit, once recounted in *Vanity Fair* magazine "one of the most embarrassing things I've ever done in public." The setting was a debate about art at the Hay Literary Festival in Wales in the mid-1990s. Gill and his debate partner, the historian Norman Stone, faced off against the British novelist Salman Rushdie and the *New Yorker* essayist Adam Gopnik. Gill and Stone contended that the pervasive influence of American culture should be resisted by the rest of the world. Rushdie and Gopnik argued for the opposite position.

Gill had the misfortune of speaking first on what he later termed a "cretin's errand." By the time he recounted the tale, the details of his opening statement had been wiped from his memory, but the gist of Gill's proclamation was that American culture was destroying the grand tradition of "drawing room, bon-mot received aesthetics"—the sort of "high art" associated with European civilization, epitomized by the classic works of Shakespeare, Michelangelo, and Mozart. Gill describes what happened next:

> After we'd proposed the damn motion, Rushdie leaned in to the microphone, paused for a moment, regarding the packed theater from those half-closed eyes, and said, soft and clear, "Be-bop-a-lula, she's my baby, / Be-bop-a-lula, I don't mean maybe. / Be-bop-a-lula, she's my baby love."

The room erupted in a torrent of cheers. When the hullabaloo finally died down, the debate was over. Rushdie's move was a "triumph of the sublime," Gill later acknowledged, because by merely reciting the nonsense lyrics of an early American rock song, Rushdie had effectively made the point that "America didn't bypass or escape civilization. It did something far more profound, far cleverer: it simply changed what civilization could be."

The expression "be-bop-a-lula" might sound trivial, but when it was delivered by a pompadoured young rock star named Gene Vincent in 1956, it meant everything that mattered to American teenagers. Implicit in Rushdie's debate tactic was the conviction that expressing the unspoken needs and desires of adolescents was a significant cultural achievement. Rock 'n' roll, thanks to its often-enigmatic lyrics, provided generations of youth with a private code for lust and rebellion that their parents couldn't hope to penetrate. "Be-bop-a-lula" spoke to their adolescent experience with an authenticity that was unprecedented in the musical arts, and Rushdie believed that this expansion of music's reach augmented and personalized the aesthetic of a generation—despite being nothing like the highbrow culture of the European intelligentsia.

Vocal music dominates the twenty-first century pop charts. When "Harlem Shake" by Baauer became a No. 1 hit in 2013, it had been thirteen years since an instrumental track had cracked the Billboard Top 10. This is surprising, considering that over the span of human (musical) history, instrumental music has always been popular. Movie themes, television themes, and even disco records featuring (instrumental) classical music made it into the Top 10 in the 1970s, including disco versions of Beethoven ("A Fifth of Beethoven" by Walter Murphy & the Big Apple Band) and a jazz-funk take on Strauss ("Also Sprach Zarathustra" by Deodato). But society's appreciation of instrumentals plunged in the '90s and shows no sign of coming back soon.

In this chapter, we delve into the most distinctively human source of musical rewards: the dimension of *lyrics*. A few other species trill out a melody or keep time with a rhythm, but only *Homo sapiens* can beget words, and words have always been a uniquely enthralling means of expressing the heart's yearnings. Lyrics can make us feel seen, heard, and understood, whether we are trying to navigate the complexities of romantic attraction for the first time or grappling with the indignities and regrets of growing old.

For many listeners, lyrics are the most resonant dimension of a record. If you're the kind of person who immediately googles the lyrics of new songs you like, you might be a lyrics-focused listener. If you can effortlessly recite the words to your favorite songs, you almost certainly are. Without the vivid imagery evoked by the words of the song, a record may earn such listeners' admiration but fail to arouse their passion.

In our own research on musical visualization, we found that the second most common type of mental imagery people see in their mind's eye when they listen to music (after personal memories) are images relating to the story in a song's lyrics. Here are some of our subjects' descriptions of what they see when they listen to their favorite records:

"I picture the lyrics of the song with people or myself."

"I imagine a story that goes along with the words."

"The story it is portraying and how it relates to my life."

"I place myself in the story that the song is conveying and internalize the emotion."

"I visualize who or what the singer is singing to and think about the relationship between the two things."

One day in the fall of 1983, just a few months after I started working with Prince, I learned firsthand how lyrics can forge an intense personal connection with listeners. A mail truck pulled up to our rehearsal warehouse in St. Louis Park, Minnesota, and delivered pallets of canvas bags stuffed to overflowing with fan mail. Bag after bag was heaped onto the gigantic fabric-cutting tables used by Prince's couturiers. Thousands of letters spilled out over the tables and onto the floor: letters in lavender envelopes, letters sprinkled with glitter, letters adorned with stickers, letters decorated in crayon, even letters bulging with gifts. This was what it looked like to have millions of "followers" before the rise of social media.

Today, a fan might take a few seconds to tweet an artist or post a comment on their Facebook page. But back then, every piece of fan mail represented a major commitment. A fan had to track down Prince's address (no easy feat before the internet); find paper, an envelope, and a stamp; write out her sentiments by hand (starting all over if there were any mistakes); decorate the envelope to help its chances of getting noticed; and physically drop the letter in a mailbox—all without any guarantee that anyone, much less Prince, would ever read it. Someone would put in that much effort only if they felt an unquenchable need to express their feeling of connection to Prince's music.

After Prince left for the night, I stayed behind with a couple of members from the Time, his protégé band. We stared in awe at the mountain of canvas bags. We didn't know what would happen to these letters, but due to their overwhelming quantity, we knew that their chances of eventually being read were somewhere between slim and none. We couldn't resist opening a few and reading them out loud.

Most appeared to be written by teenagers. Taking inspiration from Prince's *1999* album (this was before *Purple Rain* catapulted him to a whole new level of superstardom), many fans made up stories from his song titles—in an effort, I assume, to be relatable. Some wrote about the way "All the Critics Love U in New York," especially if you take a ride with a "Lady Cab Driver" on the way to meet an "International Lover" who makes you feel "Delirious." These letters had a gentle poignancy. The authors weren't asking Prince for anything. They just wanted to say that they identified with him—that his words spoke to them and for them.

This is a deeply human impulse. Lyrics can make us feel that we are experiencing life through another person's eyes, and we naturally seek out opportunities to imagine what it's like to be someone else—to think like them, move like them, and talk like them. I remember one gorgeous spring day a few months after our fan-mail adventure. I was once again at our rehearsal warehouse, adjusting mics in preparation for a recording session. The lovely weather drew the musicians, members of the Revolution and the Time, outdoors for a basketball game. I could hear them through the open bay doors of the loading dock and, wondering what it felt like to talk so differently, I whispered their banter to myself while moving mic stands: *"Boy, you bad!" "I got drawers at the crib!" "You is gone!"* Little did I know that Prince had crept up behind me and was very much amused. With a wide grin he shouted to the musicians in the courtyard:

"Hey, you guys, Susan's in here imitating y'all! You should hear her say, 'Boy, you bad!'"

I was embarrassed, but not the least bit ashamed. It was a tad mortifying to have Prince call attention to my private role-playing, but I was just doing what any lyric lover does when they sing along with records in their bedroom—trying out someone else's words and imagining what it might be like to say (or sing) them.

This is the immense power of lyrics: to enable us to momentarily become someone else.

2

As we saw in the Melody chapter, when we hear the sounds of language—whether as speech or sung lyrics—our brain quickly splits the sound's melodic content (the *intonation*) from its semantic content (the *information*). Because intonation is processed in a different brain network than information, we can consciously choose which one to focus on. Most of us have a natural preference: we may instinctively focus on the lyrics sung by the vocalists, or we may ignore the words entirely to better savor the melody.

A couple of songs from Disney movies illustrate the division. First is a song where the melody outshines the verses: one of the earliest Disney tunes, "Someday My Prince Will Come" from *Snow White and the Seven Dwarfs*, 1937. The lyrics about waiting to be rescued by a perfect man are well past their cultural expiration date, but the exquisite melody has made the song a jazz standard, performed instrumentally by Dave Brubeck, Miles Davis, and countless others. For comparison, consider "Remember Me" from *Coco*, 2017. Written by Robert Lopez and Kristen Anderson-Lopez, this Chopin-inspired melody is pretty, but the lyrical theme of staying close to our departed loved ones is what gives this song its tear-jerking power.

Though we can consciously attend to either the melody or the lyrics of a record, our brains can also unite both musical dimensions (along with rhythm and timbre) into a holistic listening experience, thereby synchronizing and amplifying the rewards we feel. Listen to how the lyrics, melody, and rhythm all seem to join hands and kipper across our consciousness in "Be Our Guest." This celebrated track, composed by Howard Ashman and Alan Menken, is from Disney's 1991 version of *Beauty and the Beast*. The lyrics are *reinforced* by the melody, and vice versa, so much so that it is difficult to imagine one without the other.

Skilled songwriters implicitly understand that the brain can run on two parallel tracks. Songwriters may opt to let lyrics and melody complement each other, as in "Be Our Guest," or to fan out the emotional content by letting each musical dimension express something different. This is what the band Train did on the track "50 Ways to Say Goodbye." The lyrics tell us that the singer is heartbroken because his girlfriend has left him. He can't bear to tell anyone, so he makes up a long list of excuses for why she's not around. Lyrically, he's in pain. Pat Monahan sings, "My heart is paralyzed," and "My pride still feels the sting / You were my everything." But melodically, the record delivers a very different message. The fast tempo keeps the mood light and upbeat. The accompaniment features telenovela-style mariachi horns and acoustic guitar for tongue-in-cheek drama. The melody suggests that Pat is just fine, despite his anguished lyrics.

The fact that musical inputs to our brains get divided into lyrics and melody before they are reunited suggests that if our ability to perceive one of these musical dimensions is lost (through brain damage, for instance), the other dimension might mitigate the impact. *Aphasia* is a neural condition commonly occurring after stroke or other brain injuries that disrupts the ability to comprehend language. In most brains, the language-processing network resides in our left temporal lobe, just above the left ear, while our melody-processing network is in the right temporal lobe, above the right ear. Thanks to this symmetry, if the speech network in the left hemisphere is damaged, the melody network in the right hemisphere can be harnessed to aid with recovery. Melodic intonation therapy (MIT) invites aphasic patients to *sing* what they have to say, thereby bypassing some of the impaired speech circuitry.

Singing is more deliberate than speaking. Because our right brain hemisphere is specialized for processing melody, it integrates incoming sounds over a longer time period than our left hemisphere does. When we slow down our speech and deliver each word at a steady pace, the sound we produce is more like singing than like speaking. The right hemisphere collectively processes *all* the words in a sentence, exactly as it does when processing individual pitches in order to organize them into melodies. Thus, singing can help aphasic patients string individual words together into a single verbal "chunk."

Our melody-processing brain network can also unify discrete phonemes, through singing, into a unified melodic word. The word "love," for example, is a single spoken syllable, but you can sing the word "love" using two or even three syllables if you assign each phoneme in the word its own note: lll-uhhh-vvv. The deliberate control of the articulation of phonemes is a useful therapeutic technique as well as an artful musical device, as we saw with Sinatra's melodic control.

When we speak, we know in advance whether we want to ask a question, declare a truth, or issue an imperative. We choose the pitch, timing, and dynamic emphasis of our vocalization so that, depending on our intention, the overall utterance has a specific melodic contour expressing the desired meaning. For example, "Did you open the *window?*" asks a different question than, "Did *you* open the window?" Research has shown that when speech-impaired patients consciously choose a specific melodic contour for what they want to say during melodic intonation therapy, it helps them regain their speech. For example, a therapist teaching an aphasic patient to say, "I would like to pet your dog" could start by first inviting the patient to select and sing a simple melody communicating the *feeling* of these words, using nonsense syllables like "la la la la la la la." Gradually, sounds that resemble the actual syllables of the words replace the nonsense syllables, perhaps like "ah wa la ta pa ya da." With practice, the patient's ability to successfully pronounce an entire sentence improves.

Rhythm also has a role to play in restoring impaired speech. Hand gestures are commonly used in both speech and singing. (Some brain scientists believe there is an ancient link between gestures and speech, one that was forged in our human ancestors before the evolution of *Homo sapiens.*) Tapping out a beat with each syllable of a desired sentence—especially tapping the fingers of the left hand, which are controlled by the right brain hemisphere—may draw upon this primal neural connection to help time our spoken words so that they are enunciated at an even pace.

One reason there are different ways to "repair" physiologically impaired speech is that the language network is heavily interlinked with many other parts of our brain, including networks responsible for vision, hearing, taste, smell, pain, pleasure, lust, love, social cognition, memories, planning, and dreaming. This promiscuous connectivity provides us with multiple therapeutic routes to support speech deficiencies.

It also empowers us to use language to express the complexities of our personal identity—and to experiment with alternative identities.

3

I was an adolescent in the summer of 1969, when America was in the midst of Vietnam-era protests and at the tail end of the race riots. External and internal changes produced a roil of questions. Would the boys in my class be sent to fight overseas? Would every social leader I admired get assassinated? Is it better to be a coed or a dropout? Why don't we have racial equality? That year the right lyric clicked into place like a seat belt for my psyche. Sly & the Family Stone's "Stand!" was on the radio:

Stand! All the things you want are real.
You have you to complete and there is no deal.

I can still remember being transfixed in the living room as I clung to the faith and optimism in Sly's voice. I felt that as long as he was putting those words out there into the world, everything was going to be all right. "Stand!" has a rhythm that drives forward with an even pressure. Cynthia and Jerry on horns build up and then answer Sly's melodic lines. Sister Rose joins in with a voice that pierces like sunlight through leaves to add shine to what he's just said. And in the midst of the last chorus, Larry breaks out into a funk groove: *Let the dancing begin!*

That year produced dozens of records I still treasure, but none remain as emotionally relevant to me as "Stand!" The reason is the lyrics.

Personal identity is a dynamic phenomenon. It changes across time and circumstances. It's in flux most dramatically when we are young and trying to figure out who we want to be—who we *could* be. Identity is both a social act (such as choosing to add "I dig!" to your discourse) and an aesthetic one (such as opting to wear hot pants but drawing the line at miniskirts). Music, and lyrics in particular, serve as a kind of encyclopedia of personas for us to explore and adopt. Lyrics provide us with a private dressing room to "try on" the words of others to see if we can fit them onto our own persona, as I did when I tried out the words spoken by members of the Time.

The identities we construct for ourselves are reflected in the things we collect and like, so much so that when we unveil a drastic change in the food we eat, the hobbies we enjoy, or the genres of music we're into, people who know us understand that something important about our identity has changed. Empirical research has shown that our conception of personal identity is linked to our musical choices. Participants in a recent study were presented with vignettes where they had to imagine drastic changes in their neighborhood, occupation, religion, or aesthetic preferences, such as imagining that after a lifetime of listening to classical music, they suddenly wanted nothing but pop. They were asked, "Would I be the same person?" and "To what extent do you think that such a dramatic change . . . would influence the relationship [with your] friends?" The results showed that the bigger the difference between musical genres—switching from listening to nothing but punk to wanting nothing but gospel, for example—the bigger the perceived shift in self-identity.

The subcultures of Deadheads, ravers, and goths are, in large part, *conceptual*: their distinctive fashions work in tandem with their distinctive music to reveal the underlying value systems embraced by these fans. This perspective was expressed by journalist John Schwartz in a 2004 *New York Times* article titled "To Know Me, Know My iPod." The opening line of this essay on new technological trends states, "Somehow, I had ended up inside of Ken's head."

Ken had sold John his iPod without bothering to delete the personal music library it held. John wrote that listening to Ken's playlist felt almost like eavesdropping, it was so intimate. He writes, "Outside his iPod, Ken is pleasant but reserved. But his selections show an unbridled feeling I had never glimpsed. . . . I know him better, and I like what I know."

Of the seven dimensions that make up your listener profile, lyrics present us with the most direct and sophisticated access port to identity. Listening to a new album can cause us to fall in love with a certain groove or an innovative sound design, but if the lyrics express ideas or values that we reject, we might reject the album, too, or at least hide it under the mattress. I recently had a casual conversation with a young journalist, in which he and I shared how much we both loved Solange's 2016 album *A Seat at the Table*. "But I would never listen to it when I'm out with my friends," he confessed. I asked why not. "Guys don't listen to albums by women when they're out with other guys," he explained. "You just don't do it." I had to hold back a smile, imagining what it would be like to retort, *I totally get it. What would the girls think of me if they found out that I liked the new John Legend album? I'd be ruined!*

I know that the beliefs of a single man don't represent the attitude of all male music lovers. Nevertheless, I recognize that his opinion harbors a grain of truth. The lyrics in the music we love are widely assumed to represent *who we are*—and for most lyric lovers, they do. Musical lyrics have special properties, absent from prose, that make them even better suited than literature for facilitating an identity swap. With their loose structure, song lyrics bear a close resemblance to inner speech or "thinking out loud." In a lyric, the speaker will often switch between talking to herself ("I need a man who'll take a chance") and talking to someone else ("Don't you wanna dance?").

In this way, pop lyrics are similar to the unself-conscious speech of young children who haven't yet reached the developmental stage where they can keep their thoughts to themselves. The author and educator Tim Murphey writes, "One might even venture to say, music and song only mean something specific to music critics, to the rest of us they merely make an abundance of sense." This idea harks back to Salman Rushdie's belief that you don't need a literal understanding of "be-bop-a-lula" as long as you *feel* what it means. Song lyrics can inhabit an emotional space inside our consciousness where rationality, specificity, and logic are wonderfully absent.

Murphey points out that if you're on a crowded street and hear someone call, "Hey, you!" you'll instinctively turn around, though usually you'll discover that the shouter was addressing someone else. When you're listening to the lyrics of a record, however, your instinctive assumption that the word "you" refers to *you* never gets challenged. Your mind automatically puts yourself in the "I" or "you" position of a song's lyrics, even if sometimes you'd prefer not to.

A young drummer I worked with in the mid-1990s was going through a messy and painful romantic breakup. One morning, he came into the studio and wailed, *"Why is everyone singing songs about me?!"* Listening to the radio and hearing potent love songs like "Kiss from a Rose" by Seal and "You Are Not Alone" by Michael Jackson was too much for him to bear during that painful time.

Skilled songwriters exploit the fact that lyrics can evoke the experience of inner speech. Murphey also notes that pop music lyricists usually leave the time, place, and persons unnamed, and this makes it easier for listeners to fit the words into their own personal story. Modern song lyrics typically don't describe precise events but are in-the-moment descriptions of ongoing emotions or circumstances. Just as a good con man includes just enough details in his lie to make the con believable, songwriters deploy just enough detail to make you believe in the authenticity of the song's fictional realm.

"This Is What Makes Us Girls" by Lana Del Rey works this way for me. The vivid images that Del Rey plants with lines about being a teenager "table dancin' at the local dive" and her best friend's "mascara running down her little Bambi eyes" and the boys who "whistle 'Hi, hi'" and drinking "Pabst Blue Ribbon on ice" let me picture things that I haven't done personally but (having once been a teenager) know all about. I cherish the song nearly as much as a real memory because Del Rey allows me to see and feel it.

Talented songwriters recognize that listeners' imaginations work alongside their literal perception when they are enjoying a record. Even as we process the words, our minds are spinning out imagery and memories and emotions. With "This Is What Makes Us Girls," I see wet asphalt streaked with neon-colored light from a liquor store at night and a parapet along a shallow river. I hear the sounds of kids laughing and running and yelling, and I smell cigarettes, perfume, bug spray, and pot in the summer air. None of that is in the song. Del Rey's lyrics help my imagination fill in the blanks and, over the course of four minutes, I am *living* there.

Writers intentionally use ambiguity like this to provide a more engaging experience, inviting listeners to concoct their own story around the missing parts. As we saw with abstract art in the Realism chapter, ambiguity lets the viewer explore several possibilities in a search for meaning. It can feel satisfying to believe that you have figured out what a song is about, just as it can be rewarding to detect a pattern in an abstract painting. These feelings can endow you with a sense of belonging to the community that "gets it."

Don McLean's 1971 "American Pie" is an eight-and-a-half-minute-long acoustic opus that was ranked No. 5 on the Recording Industry Association of America's "Top 365 Songs of the Twentieth Century." The song is famous for its enigmatic lyrics, such as:

When the jester sang for the king and queen
In a coat he borrowed from James Dean

McLean wasn't the first lyricist whose songs were narrative poems—Bob Dylan had been doing that for years—but "American Pie" caught the imagination of more listeners and became "an anthem for a generation," many of whose members memorized every line. The colorful lyrics were cryptic and earnest enough to seem as though they expressed the root cause of social and political feuds. Even today, fan sites pore over the perceived symbolism of every metaphor in the song. For four decades, McLean steadfastly refused to say what he intended the song to be about. When asked what the verses meant, he often replied, "It means I'll never have to work again."

Then, in 2015, while selling his original handwritten manuscript for the lyrics of "American Pie" at auction, McLean revealed in a note printed in the Christie's Auctions Catalog that the inscrutable verses were "not a parlor game." In an interview for the catalog McLean went on to say, "Basically in 'American Pie,' things are heading in the wrong direction. It is becoming less idyllic. I don't know whether you consider that wrong or right but it is a morality song in a sense." It was probably a good idea for McLean to avoid explaining his lyrics all those years, for his explanation is a lot less interesting than many fans' interpretations.

I confronted the power of lyrical ambiguity years ago when I had the good fortune to engineer for singer-songwriter Edie Brickell, who was (and still is) Mrs. Paul Simon. One of Simon's most popular and enduring songs—"Me and Julio Down by the Schoolyard"— describes how "mama pajama . . . ran to the police station" after "seein' me and Julio down by the schoolyard" doing something— what?—that "was against the law." When I told my manager at the time that I had an opportunity to work with Edie, the first thing he said to me was, "Oh please, ask her to ask Paul what he and Julio were doing down by the schoolyard!"

Knowing that lyrics should not be taken literally, I never asked. Some lyrics raise intriguing questions, and great writers like Paul Simon know that not every question must be answered.

4

Lyrics expose us to songwriters' worlds—to their value systems and perspectives. And when lyrics repeatedly ring the bell of self-recognition, it can cause us to feel as if we know the writer personally. But songwriters have their own identities, with their own private aspirations and fears that they never share with listeners. They often write lyrics that are complete fiction.

Most of the time, listeners have no way of knowing the extent to which the words of a song represent the writer's lived experiences. Lyricists aren't always willing to share what inspired a song or a clever line. In the case of posthumous artists, record producers and band members are sometimes asked in interviews to "spill the beans" and say who a song was written about. Usually, the honest answer is that there was no single muse. An amorous encounter, a breakup, an epiphany, a disappointment, a strong hunch—anything can be a starting point for a lyric. But once the writing journey begins, creative invention often takes over.

For record producers, make-believe is often preferable to the sincere yet boring "diary entries" that novice lyricists are inclined to come up with. A blend of fact and fiction is more likely to resemble the listener's own experiences. As we saw in the Authenticity chapter, a sense of authenticity can be leveraged to establish a connection with listeners, though it doesn't follow that every word that helps forge that bond must be true. My younger brother John once told me how much he loved a breakup song I co-produced: "Call and Answer" on Barenaked Ladies' *Stunt* album. This was a little puzzling because John has the enviable experience of being in a decades-long marriage with his one-and-only love, yet he identified strongly with the song's lyrics. So I was touched when he said, "I've never broken up with anybody, *but if I did*, that would be the song I'd want to sing."

For a songwriter, providing a listener with words for an imagined experience that feels utterly real is as good as it gets, so I shared my brother's reaction with Steven Page, who co-wrote the song with Stephen Duffy. Steven flashed a broad, amused grin, but he looked a little embarrassed. The song was complete fiction. He and Duffy made it up.

At the close of a long recording session at Sunset Sound in Los Angeles, I encountered another example of the mismatch between an artist's true identity and his identity as expressed in his lyrics. I was packing up Prince's tapes and equipment to ship home. A young staffer was helping me while asking questions about Prince, whom he had never met. I answered his questions, then asked which artist he liked the most. Bruce Springsteen, he replied. I asked why, expecting to hear something about the Boss's songwriting style. "Well," he said, "I get the feeling that if Bruce met me, he'd like me. He'd sit down and have a beer with me. But I get the feeling from Prince that he'd steal my girlfriend." Given Prince's lyrics, the staffer's presumption was understandable, if inaccurate.

Many Prince songs get right to the point: "I wanna be the only one U come for," "You've got a wonderful ass," and "Let me touch your body, Baby, let me feel U up." These lyrics are more direct than Springsteen's "I got a bad desire," and "This gun's for hire, / Even if we're just dancing in the dark," though the subtext is the same. In reality, Prince was scrupulous in his respect for personal relationships. It made me a little sad when listeners' impressions of him didn't jibe with the man I knew.

When you embrace a songwriter's lyrical identity, your own identity can feel expanded or transformed. Soul music gave me a window into the isolation caused by racism and occasionally helped me arrive at the same point of view as that of someone whose life experiences were very different from my own. The day I brought home the new Public Enemy album *Fear of a Black Planet*, I blasted the song "Fight the Power" and listened raptly. The track's arrangement, sound design, performances, and engineering were innovative and thrilling. I had never heard anything quite like it. But when these lyrics blared out, I sprang up with a raised fist and shouted "YES!" to the empty room:

You see straight up racist the sucker was, simple and plain
Muthafuck him and John Wayne.

Only once before had I heard anyone put down movie icon John Wayne. For an earlier generation he had been a larger-than-life American hero. Wayne was held up to boys in the 1940s and '50s as a symbol of American manhood, though his movie roles often featured him slaughtering Indigenous Americans. It turned out I wasn't alone in questioning whether John Wayne deserved hero status. Listening to "Fight the Power" and Chuck D's body of lyrical work, I felt a fellowship with him—a man with whom I had little in common in cultural origins but plenty in common in the conclusions we reached.

Sometimes artists explicitly craft a fake persona: a fictional character expressing thoughts and feelings that fit an alternative ego. Fans find much to love in these characters, especially when they adopt a unique look, as did '70s glam rock band Kiss and George Clinton's Dr. Funkenstein. Beyoncé's alter ego Sasha Fierce appeared on one album to let Beyoncé try out electropop music, which didn't perfectly fit with the star we know. Garth Brooks created the alter ego "Chris Gaines," sporting black eyeliner and long hair, to fully immerse himself in rock music without having to abandon his country base. Nicki Minaj has created a number of alter egos, but the reported fan favorite is Roman Zolanski, a gay man from London. The Ramones—all unrelated in real life—adopted a single identity expressed in four bodies: Joey Ramone, Dee Dee Ramone, Johnny Ramone, and Tommy Ramone.

More recently, Elizabeth Woolridge Grant created the character

Lana Del Rey to more fully express the lyrical themes that appeal to her. One of her earliest appearances as "Lana Del Rey" was her 2012 performance on *Saturday Night Live*. Some viewers mocked her on social media, with their descriptions of her performance ranging from "unconfident" to "unhinged." I didn't see it that way. It looked to me as though she was working on the Lana character but it hadn't yet become fully integrated with Elizabeth the artist. Her performance was reminiscent of David Jones's early days when his "David Bowie" character was first taking shape, Reginald Dwight's early appearances as "Elton John," and Herman Blount's "Sun Ra." Each of these not-so-secret identities came to life over time through affected mannerisms, stage costumes, and, crucially, through the lyrics of the artists' songs.

Lyrics is the only musical dimension in our listener profile that offers a two-way connection between music makers and audiences. Most listeners can't or won't respond to an artist's song by writing a new melody or rhythm. But anybody can talk back to an artist. As fans learn about songwriters from their lyrics, so do songwriters learn about their fans from their letters, online comments, or conversations. Songwriters are aware that what they release into the world is a mere "slice of me"—an incomplete and often distorted view of their beliefs, thoughts, feelings, interests, and dreams. It takes empathy and a bit of wisdom for a songwriter to realize that this "slice" is what fans expect to encounter when they meet him.

Years ago, my friend Tim Bruckner, a sculptor, had a once-in-a-lifetime opportunity to meet with his hero, John Lennon, and discuss artwork for an upcoming project. Tim felt as nervous as a rat in a coffee can as he waited in the kitchen at breakfast time, portfolio under his arm, trying to figure out what he would say to the legendary figure. A housekeeper set cereal, a bowl, and some milk on the table in front of Tim. Lennon walked in and sat down, served himself, and placed the box of cornflakes between them. Tim watched expectantly as his idol poured milk into his own cereal bowl. Then

Lennon looked up with a soft grin and in a deep Liverpudlian accent broke the tension by saying, "Ah! Sittin' on a cornflake!" Tim's jaw must have dropped, because Lennon let his own jaw fall, then motioned pushing it back up with a finger.

In quoting his own lyrics to the young guest, Lennon was saying, *I know what you see when you look at me. We can meet on the street where you recognize me.*

5

The dual mental engines for lyrics and melody lets musical text be nonsensical and still connect with listeners, as we saw with "Be-Bop-a-Lula." Musicologists distinguish between two kinds of nonsense lyrics: syllabic and propositional. Syllabic nonsense is made of vocal sounds that have no individual meaning: they sound good and seem to fit together, like "be-bop-a-lula." Syllabic nonsense has made its way into hit records in every decade of popular music: "Ooby Dooby" by Roy Orbison, "Goo goo g'joob" in the Beatles' "I Am the Walrus," "De Do Do Do De Da Da Da" by the Police, "MMMbop" by Hanson, and "Bom Bidi Bom" by Nick Jonas and Nicki Minaj. To some academics, the meaningless sounds may signify rebellion or subversion, but to the fans of these songs, they're an invitation to a private club.

Occasionally, songwriters use nonsense lyrics as placeholders before writing more sensible lyrics. One colorful example is found in the Irish rock band U2. The lead singer, Bono, is famous for "scatting" in rehearsal in search of syllables that work well with the rhythm and chord progression. "We call it Bongolese," says Andy Barlow, the British record producer who contributed to U2's *Songs of Experience*. "He'll basically make up words about the view or the cup of coffee he's drinking—just pure channeling—and from that we would find what feels good and piece it together." Every now and then, nonsense syllables that were originally lyrical substitutes seem so well integrated with the other elements that after a few rehearsals, there seems to be no good reason to replace them.

The second kind of nonsense lyrics is propositional. In propositional nonsense, the individual words are legitimate and meaningful, but when the words are strung together, the verses don't obey conventional grammar or may refer to things that are illogical or surreal. My coauthor is fond of "I Before E Except After C," a record by the 1980s electropop band Yaz that features band members (and the producer's mother) reading from an audio-equipment technical manual and having their spoken words rearranged into nonsensical verbal patterns.

A more poetic example of propositional nonsense is found in the Beach Boys' masterpiece "Surf's Up," said by some music observers to portend Brian Wilson's still-undiagnosed mental illness. The record gives me chills. Despite its oblique word pairings (such as "Dove nested towers the hour was" and "The young and often spring you gave"), I feel like I know what it is expressing. For me, "Surf's Up" portrays a great songwriter struggling to maintain a foothold on life, and I find this vulnerability deeply moving.

6

For those listeners who obtain musical rewards from lyrics, there is a good chance that you first recognized yourself in songs when you were young. Adolescence and music go together like ice cream and hot fudge. That sweet and volatile period between childhood and adulthood, best characterized by our quest for identity, can be comforted and indulged by the verbal messages in music.

Today's adolescent enjoys a rich endowment of music—collaborative, interactive, publicly shared, privately enjoyed—to assist with the task of figuring out what to think, say, do, or be. For most teens, the lyrics of pop music address concepts and situations that might not be fully unpacked in school or at home, such as sex, drugs, depression, anxiety, social rejection, romance, and violence. A popular song may be a teen's chief source of intel on serious but confusing matters. Bruce Springsteen captured this sentiment in a line from "No Surrender":

We learned more from a three-minute record, baby,
Than we ever learned in school

A meta-analysis published in the journal *Psychology of Music* reported on lyrical themes in American Top 40 pop songs from 1960 to 2010. Ever since the 1960s, the predominant themes in pop music lyrics have been romance and sexual relations, although as the decades go by, the scales have tipped in favor of more sex and less romance. Lyrical depictions of lifestyles that include dancing, substance use and abuse, and status or wealth increased sharply after the turn of the millennium. Lyrics about social changes, religion, personal identity, family, and friends represent a stable share of themes: they have trended neither up nor down over the years. Violent lyrical imagery, however, showed a sharp increase in the 1990s. Rap music witnessed the most dramatic increase in representations of violence (from 27 percent to 60 percent) between the years 1979 and 1997. Lyrics in the late '90s often portrayed violence in a positive light, associating it with strength, masculinity, and wealth.

But many of the world's most popular songs feature lyrics that capture the social anxiety and isolation of youth. One of the most quintessential expressions of the emotional island of adolescence is the Beach Boys' 1963 hit "In My Room," written by Brian Wilson and his friend Gary Usher when Wilson was barely out of his teens. Like so many early Beach Boys songs, it has lyrics so simple that only a young person would write them and only a young (or young-at-heart) listener can fully appreciate them.

Wilson was agoraphobic for periods of his life. Although the song was written before the disorder set in, he and Usher tapped the deep feeling that your room is your sanctuary, the one and only place where you can both *be* yourself and find the courage to *change* yourself. Verses such as "Now it's dark and I'm alone / But I won't be afraid" are pure, distilled emotional truth, the kind we openly express when we're young but suppress or mistrust in adulthood.

Steve Perry—the vocalist and songwriter of Journey fame—talked about the solace of "In My Room" with *Rolling Stone*: "This was an anthem to my teenage isolation. I just wanted to be left alone in my room, where I could find peace of mind and play music." It's easy to imagine that Perry and his bandmate Jonathan Cain were inspired by the purity of "In My Room" when they penned an idealistic anthem of their own: "Don't Stop Believin'." Jim Axelrod of CBS news writes of the tune: "It may have started as solely Journey's song—but it belongs to all of us now."

7

Scientists have used the psychological concept of *self-congruity* to posit that music listeners prefer artists whose personality traits (gleaned through lyrics and appearance) are assumed to match their own. This idea was supported by a large-scale study published in 2020. The data added to the body of evidence examining music and social identity. Music listeners report that they choose artists and records that reflect and reinforce their personality traits and address their psychological or social needs.

But psychological needs change as we go through life. Many lyrics that appealed to us in our teens won't feel as urgent or compelling when we are older. Likewise, a song we ignored in youth may take on importance as we age.

In early 2021, Olivia Rodrigo's smash hit "drivers license" broke

Spotify's record for the most non-holiday song streams in one day. The melody is pretty and the vocals sincere, yet on the page the lyrics are undeveloped and repetitive. Without the weight of life and experience to add layers of perspective to their point of view, the song's verses repeat the same thing over and over with slightly different words, accurately reflecting the naïve experience of a young relationship.

The adolescent mind is astoundingly sensitive to the minds of its peers. Teens place a far greater weight on social acceptance than do children and adults. In a neuroimaging study, when adults were asked to imagine "what others think of you" and then asked to imagine "what you think of yourself," the resulting images revealed activity in two distinct regions of the brain that only partially overlapped. But when teens imagined the two perspectives of their self, the two regions of activity showed more significant overlap. Neurally, what teens think of themselves is almost identical to their impression of what others think about them. Effective social cognition relies on self-knowledge and self-awareness, and both are under construction throughout adolescence. The lyrics of a smash hit by a teenage songwriter are less likely to be critically analyzed by other teens and more likely to be adopted as a blueprint and comfort blanket: "This is how to think, speak, and act."

As we leave youth behind and enter adulthood, we face new challenges. One of the biggest is choosing the right mate. Our growing stockpile of experiences helps us to recognize more sophisticated meanings in a song's lyrics. Carrie Underwood's "Before He Cheats" demands a more mature perspective than most teens have yet acquired in order to understand the implication of lines like "He's probably buyin' her some fruity little drink, / 'Cause she can't shoot whiskey." These words imply who the singer is (she *can* shoot whiskey) and why she took "a Louisville slugger to both headlights." Those experiencing the pain of infidelity may be tempted to assign blame entirely to themselves. But if we imagine someone as lovely as

Carrie Underwood struggling with the same issue, it can comfort us to realize that cheating is a human problem and not just a personal one. Adult breakup songs like the Tammy Wynette classic "D-I-V-O-R-C-E" are best appreciated by those for whom this issue is concrete reality, or at least a looming possibility.

Another song about romance that is perhaps most cherished by older listeners: "Always On My Mind," made famous by Willie Nelson. The lyrics express the complexities and regrets of love when viewed over the fullness of time. Lines like "Little things I should have said and done, / I just never took the time" don't mean as much to young listeners, who prefer explicitness over subtlety and have plenty of time to right their wrongs. This song's lyrics take on greater significance for the listener who has lost that chance.

Lyrics serve our social lives by stirring up our memories. Indeed, autobiographical memories are the single most common form of visualization that listeners report experiencing while listening to their favorite music. Many people enjoy reliving scenes from their past, and cite the desire for reminiscence as their main reason for listening to music.

When we retrieve personally relevant moments from our long-term memory bank, we engage the medial prefrontal cortex—a brain region associated with social cognition. Autobiographical memories can evoke a pleasant feeling of nostalgia, followed by relief from loneliness. By facilitating the natural process of memory retrieval, listening to music can be an effective social companion during times of stress or isolation.

The rewards of nostalgia are also driven by the shifting contexts of culture. A newly released record is heard in relation to all the other popular records of the day. When a particular record came out, it may not have thrilled you. But if you hear it again years later, you may automatically recall the time and place you first heard it, thereby triggering a surprising warmth toward the auditory memory from your past. Your mind is less critical of the specific musical qualities of the track, and more appreciative of the milieu in which you originally listened to it.

Lyrics animate your awareness of your private self like no other dimension of your listener profile. Simple expressions of effort, like "Running up that hill," or origins, like "We come from the land of the ice and snow," or dreams, like "I know I can be what I wanna be" put concrete words to vague feelings, anchoring our identity in poetry. Words that express our secret fears and longings resonate most powerfully with us. Whether you are drawn to hallucinogenic scribblings inspired by drug trips, the solemn paeans of grand opera, the sultry ballads of seduction, or gritty "slice of life" street poems, the lyrics you revel in the most reflect who you are, what you value, and—every now and then—who you'd like to be.

Chills from Music

When many people listen to music, they experience chills, a physiological phenomenon whose formal moniker is "piloerection." Chills are an automatic response associated with the capacity for empathy. Women and musicians are more likely to get chills while listening to music than men and non-musicians.

Chills can arise when some aspect of music—such as an unexpected harmony, a sudden crescendo, or a piercing high note—sounds like a child or animal in distress. Some researchers contend that chills may be the physiological manifestation of the experience of separation from our loved ones. Others link chills to auditory "looming"—where a sound is perceived as an approaching threat. Once we realize that the chill-inducing music is not actually a danger signal, pleasure can follow. This is similar to an amusement park ride supplying us with the welcome feeling of "pseudo-danger."

Record makers try to induce a chill response by boosting the loudness and brightness of an instrument or voice, but these techniques do not work for every type of music, nor for every listener. When it comes to music and emotion, variability is the rule.

CHAPTER 6

RHYTHM

This Is What Music Moves Like

◈

Don't listen to the music on the record.
Listen to the rhythm.

—*Sam Phillips, record producer*

1

LET'S KEEP ROLLING WITH OUR RECORD PULL. IT'S TIME to turn our attention to my favorite musical dimension: rhythm. I selected the following track to help you focus on how your body *feels* the rhythm. It doesn't matter if you like the record or not: the point here is to discover how you naturally synchronize your body to a song. As you listen, pay attention to the beats in the rhythm that you feel most strongly—and *tap along to those beats with your fingers*.

Go ahead and listen to "Stoned and Starving" by Parquet Courts.

This record features the most common time signature in contemporary Western music: 4/4 time. This means that each structural unit of the song (a "bar") contains four beats (in formal terms, four quarter notes): one–two–three–four, one–two–three–four. In "Stoned and Starving," the four beats in each bar are alternately played by the kick drum and snare drum: kick–snare–kick–snare, kick–snare–kick–snare. The rhythm guitar and hi-hat play eighth notes, meaning that they play along with the kick and snare but also play an additional beat in between: one–and–two–and–three–and–four–and, one–and–two–and–three–and–four–and.

So then. Which beats did you instinctively tap your fingers to?

Perhaps you tapped along with the snare, twice in every bar: one–TWO–three–FOUR, one–TWO–three–FOUR. If so, you felt the rhythm on the *backbeat* (also called the *upbeat*). Or you may have synchronized your taps with the kick drum: ONE–two–THREE–four, ONE–two–THREE–four. If so, you felt the rhythm on the *downbeat*. Or you may have tapped four times in every bar, hitting every quarter note: ONE–TWO–THREE–FOUR, ONE–TWO–THREE–FOUR. Or maybe you tapped along with every hit of the hi-hat, matching the fastest explicit beat on the record: ONE–AND–TWO–AND–THREE–AND–FOUR–AND, ONE–AND–TWO–AND–THREE–AND–FOUR–AND.

Personally, I feel this record's rhythm on the backbeat. My coauthor, on the other hand, feels the rhythm on the downbeat. The varied ways that different listeners perceive the rhythm on a record illustrate this chapter's most important lesson: *your own experience of rhythm is almost entirely subjective.*

If you ask a dozen people to hum the melody or quote the lyrics of a well-known song, you'll get a dozen (nearly) identical responses. But if you ask a dozen people how they dance to the song, you will see that not everybody's interpretation of its rhythm agrees. Though a song's time signature and tempo are objective properties that can be accurately notated on a score, its *rhythm* is a psychological property. Often, there is no agreed-upon answer to the question, Where do the "true" beats land in this rhythm?

Music theorists use the word "tactus" to refer to your own personal interpretation of where you feel the beat on a particular record. The word "tactus" is related to "tactile": rhythm seems to *touch* our body. All of us experience the feeling of silk as smooth and burlap as rough, but the tactile experience of more complex textures, like denim, corduroy, or jersey knits, varies from person to person. Depending on our personal sense of touch, we may feel these textures as smoother or rougher than others feel them. (Suede feels soft to me, while my coauthor insists that it is moderately rough.) Likewise, for records with simple, steady rhythms, most listeners will share a common tactus, but for more complex rhythms we may each feel the beat in our own way.

The way you move in time to music is distinctive, a reflection of the deeply personal relationship between music, your body, and the unique wiring of your brain. A listener tapping on the downbeat in

"Stoned and Starving" feels a different tactus than someone tapping to all four beats in a bar, and they both feel a different tactus than someone who taps along to the eight hits of the hi-hat. We can observe this variability in rhythm perception in laboratory experiments that examine people's ability to keep time with a beat. When listeners are asked to tap along with the unvarying, evenly spaced ticks of a metronome, we all experience the same tactus: everyone taps once per tick. The metronome provides an "objective" beat that we can all agree upon. The difference between the objective beat of a metronome and the subjective beat we perceive on "Stoned and Starving" has implications for the rhythm dimension of our listener profile—implications that record makers eagerly exploit.

In the recording studio, rhythm is expressed through two types of performance gestures: *accents*—loud beats versus soft beats—and *timing*. Accents provide us with mental clues suggesting how the music is organized into bars, allowing us to subliminally count along and anticipate changes in the musical structure, as we learned in the Novelty chapter. Accents are also used to suggest the type of dance that suits a record best. Your sweet spots on the dimension of rhythm are most likely represented by those records that place the accents precisely where they match the way you like to move. Not all music features *accented rhythms*, however. Dance pop, EDM, and techno feature *metronomic rhythms*—uniform, unaccented rhythms. Yet even if a record contains no accents, each listener still identifies the tactus that best suits her body.

Our fondness for the tactus we perceive on a record is also influenced by the *timing* of the band's performance gestures, especially on realistic records where the drummer's body is controlling the groove. Drummers opting to record while listening to a "click track" (a metronomic *tick* used to establish tempo) rely on an external source of time to make sure that the tempo remains steady from the beginning of the song to the very end. This often has advantages. It lets listeners predict acoustic events with greater accuracy—we can know *exactly* when the next beat will arrive—thereby allowing tension to build slowly over the full length of the song, rather than section by section. But a click track can also impose unwanted constraints.

Much of the pleasure we derive from music listening comes from its tension and release—the push and pull of building up anticipation and letting it go. Tension and release can be expressed by varying the timing as well as the intensity of music's beats. Typically, drummers induce tension by speeding up as they approach the end of a section, then giving us a sigh of release, subtly reducing their speed as the next section gets underway. A click track might eliminate the elasticity (the push and pull over time) in a rhythm. But even if the song's tempo is steady, producers can still introduce elasticity by asking the drummer to put his snare "a little bit behind the beat" or to "push the hi-hat." The great session drummer Jim Keltner, known for his work with Bob Dylan, George Harrison, and Eric Clapton, has the remarkable skill of timing his tom fills in such a way that you don't think he could possibly make it to the down-beat on time—yet he always does. This performance gesture adds a syrupy elasticity to parts of the track. Charlie Watts—drummer for the Rolling Stones—has been described as "playing with a minimum of motion, often slightly behind the beat," which produced "a barely perceptible but inimitable rhythmic drag." Small manipulations like these are what make some records feel just right for your body.

2

In the days before digital handclaps, record makers recorded claps the old-fashioned way: by gathering a group of people around a mic and having them clap together on the beat. When Prince and his crew recorded at home in Minneapolis, we always had musicians around for recording "clap tracks." But when we worked at Sunset

Sound studio in Los Angeles, we sometimes needed to recruit errand runners or folks from the front desk.

When Prince called for "soul claps," it meant to clap double-time, on every eighth-note: ONE–AND–TWO–AND–THREE–AND–FOUR–AND. He would often call for them onstage during funk songs to get some audience participation. (You can hear him shout, "Are you ready, Paris?! Soul claps!" at the 5:15 mark on "It's Gonna Be a Beautiful Night" from the *Sign o' the Times* album.) One day at Sunset Sound it was time to get some soul claps recorded. Prince, a couple of band members, the assistant engineer, and a young female staffer from the front desk put on headphones, gathered around a mic, and started clapping.

Everything started off well enough. As the song progressed, however, the staffer's claps got further and further off the beat. Puzzled, I stopped the tape, rolled back, and started over. Again, she began well enough, but as the song progressed, her claps grew more and more out of time. Prince signaled me to stop. With an expressionless face, he looked at the woman, raised his arm, and pointed at the door. This was my first encounter with *beat deafness*.

Just as there are people who can't carry a tune (tone deafness), there are also people with "two left feet" who can't follow a beat. Beat-deaf individuals are a valuable resource for neuroscientists. Every mechanic knows that one of the best ways to learn how cars work is to study cars that *aren't* working. One way to understand how rhythm perception works is to study people whose rhythm perception isn't working—folks who cannot dance to the beat, no matter how hard they try.

A coterie of leading music-cognition researchers, led by psychologists Jessica Phillips-Silver and Isabelle Peretz, conducted a series of experiments on beat deafness in 2011. The researchers posted an ad soliciting volunteers who could not keep time to music. To the scientists' dismay, most respondents were not truly beat-deaf. Though the respondents honestly believed they had no rhythm, when they were tested in the lab, it turned out that they exhibited at least a modest tactus. Every volunteer could reliably tap along in time with the perceived beat of a song. Fortunately for the researchers, there was one exception: a twenty-three-year-old student named Mathieu.

Mathieu loved music. He had even taken music and dance lessons. However, he confessed that finding the rhythm had always been difficult for him. He had no trouble with pitch or melodies. Mathieu easily passed five of the six tests in the Montreal Battery of Evaluation of Amusia (MBEA), a widely adopted measure of tone deafness. Most of these tests present a pair of melodies and ask the listener to say if they are the same or different. But there was one test on the MBEA that proved insurmountable for Mathieu, even though most tone-deaf listeners find it easy: listening to short piano pieces and deciding whether the rhythm is a march (ONE–two, ONE–two . . .) or a waltz (ONE–two–three, ONE–two–three . . .).

After taking the MBEA, Mathieu and thirty-three other volunteers were each tested on their ability to move along to three different sources of rhythm: a metronome, a live dancer, and a merengue record. Mathieu was in top form when synchronizing to the metronome; he had little difficulty keeping time with its steady clicks. Nor did he exhibit problems synchronizing his movements with the *visually* perceived movements of the dancer. But everything fell apart when Mathieu listened to the merengue record "Suavemente" by GRAMMY-winning artist Elvis Crespo, chosen by the Montreal researchers for its strong binary rhythm (a 2/4 time signature, with two beats per bar).

In his attempts to find the rhythm, Mathieu tended to move *in between* the beats rather than on the beats. When the researchers brought the dancer back to move in time with the song, Mathieu was once again able to stay in sync. But as soon as the dancer stopped, forcing Mathieu to rely on his ears alone, he again fell out of time.

Despite listening to intensely rhythmic music, Mathieu was unable to experience a tactus.

Mathieu was also tested on his ability to synchronize with other kinds of music, including swing, techno, Egyptian percussion, world, dance pop, rock dance, and lounge dance. He did moderately well with dance pop and techno—two genres dominated by computer-programmed, metronomic rhythms—but he could not stay in time with any of the other genres.

Cases of beat deafness are especially odd because few behaviors come more naturally to humans than perceiving a beat. Infants as young as five months old spontaneously move to rhythmic music, although they lack the motor skills to keep in time. Most toddlers can't resist the urge to march and clap to music but must wait until age four or so before their bodies are coordinated enough to allow them to truly synchronize with a groove. This suggests that *Homo sapiens* are born with the neural infrastructure for extracting a subjective rhythm from regularly timed events, though it takes a few years for our motor-control circuitry to catch up with our brain's perceptual ability.

By the time we reach adulthood, most of us have highly competent rhythm perception. We can even convert complex musical structures featuring multiple layers of beats (called "trees of time" by biologist and cognitive scientist Tecumseh Fitch) into a dance-able rhythm. For example, listen carefully to the opening bars of Missy Elliott's 2001 infectious dance hit "Get Ur Freak On." You'll hear three different sources of percussive rhythm. Most prominently, there's the much-imitated tumbi, a single-stringed instrument from India that has been twanging from car windows every summer since producer Timbaland first popularized it in the West on this very record. Next, there's the low digital kick drum from the classic Roland TR-808 drum machine. It plays both down-beats and backbeats. Finally, there are tablas (Indian hand drums) tapping out a string of eighth, sixteenth, and thirty-second notes.

In total, there are no fewer than thirty-four percussive hits in each bar. This is a frenetic tempest of percussion compared to the pristine simplicity of a "four-on-the-floor" beat (where the kick drum plays on every quarter note, as in most electronic dance music). The international popularity of "Get Ur Freak On" demonstrates that humans easily extract a groove from this storm of beats and dance to it with ease.

Most humans, that is. What accounted for Mathieu's lack of rhythm? How was he able to keep time with a metronome, yet get tripped up by the vividly accented beats in a merengue? The answer to this riddle was discovered in a most unlikely place.

3

"Humans are the only species to spontaneously synchronize to the beat of music," Aniruddh Patel, a renowned music-cognition researcher, wrote in his 2007 book *Music, Language, and the Brain*. Patel's proclamation did not provoke the slightest whiff of controversy. At the time, it was believed that the ability to extract rhythm from a highly complex acoustic pattern (such as in "Get Ur Freak On") required neural circuitry of such sophistication that only the brain of *Homo sapiens* had attained it.

In the decade prior to Patel's 2007 assertion, theories of rhythm perception usually made no distinction between *entrainment* (moving in time to a metronomic click) and *beat induction* (moving in time to an accented rhythm). Both mental skills were believed to be embodied within timekeeper neural circuitry that listened for regularly timed events. In essence, scientists believed that our brain's internal clock, a well-studied neural system that synchronizes with daily physiological cycles (such as the sleep-wake cycle), also synchronized with the rhythms expressed in music. Conventional wisdom held that some animals were capable of entrainment (synchronizing with a metronome), but only humans were capable of beat induction (such as synchronizing with "Get Ur Freak On").

Everything changed when Patel laid eyes on Snowball.

Patel's graduate students introduced him to a series of YouTube videos starring a sulfur-crested cockatoo standing just over a foot tall. Snowball wasn't merely dancing in these videos. The bird was grooving with all the flair and passion of a mohawked B-boy at a street jam. In the earliest video, Snowball is perched regally on the back of a club chair in a modest suburban home. The Backstreet Boys' "Everybody (Backstreet's Back)" plays. When the drums disappear from the music for a bar or two (called a *breakdown*), Snowball stops moving . . . only to pick it back up again when the groove resumes. Just like humans on the dance floor, Snowball frequently changes it up: he swings to the right, swings to the left, then lets loose during the loudest parts of the music.

Other videos soon followed, featuring other dance tracks: Queen, Michael Jackson, even German polkas. Snowball grooved to them all. It wasn't long before Madison Avenue called. Snowball starred in a Taco Bell commercial, doing his best to synchronize to the less-upbeat Rupert Holmes song "Escape (The Piña Colada Song)." The best-selling nature author Sy Montgomery penned Snowball's story in a children's book. Soon the little cockatoo became the most celebrated non-human dancer on earth.

Before Snowball's videos appeared, scientists knew that some animals could *imitate* human movements, so that if you started bouncing your head to music, for instance, a seal or an elephant might bounce along with you. But the idea of an animal spontaneously finding a personal tactus and choosing dance moves according to its own private notion of what the music called for was, for Patel, as preposterous as "a dog reading a newspaper out loud." Thus, Patel approached Snowball's dance videos with his scientific skepticism dialed all the way up.

Snowball's caregiver was Irena Schulz, the director of the Bird

Lovers Only rescue center in Dyer, Indiana. Patel knew that Schulz could be signaling to the bird from off camera. Or perhaps Snowball was imitating an unseen human dancer. Or maybe Schulz had painstakingly trained Snowball to follow a complex routine in return for rewards. To learn the truth, Patel contacted Schulz, who invited him to come visit the rescue center so that he could scrutinize Snowball's behavior firsthand.

Schulz recounted to Patel the story of how Snowball's previous caregiver had to give up the "teenaged" bird to the shelter. As the young man left, he casually mentioned that the little cockatoo loved to dance. When Schulz and her husband first saw Snowball gyrating along with music, they captured the performance on video and posted it online. As Patel soon learned, Snowball had not been trained. If he was in earshot of a dance track, Snowball could not resist moving in sync according to his own notion of the rhythm. He did not receive treats or any human-managed reward for his performances. His enthusiasm was self-generated. Apparently, the sound of bass and drums compelled Snowball to dance for the same reason we do: *because it feels good.*

In a single video, Snowball demolished the notion that tactus was a humans-only affair. But Snowball was even more remarkable than the researchers had ever dreamed. The cockatoo had developed a repertoire of no fewer than *fourteen* distinct dance moves. With his body bouncing, foot tapping, side-to-side steps, a fancy foot-lift downswing, and a pose that would impress any *Vogue* reader, Snowball deploys moves suited to the particular beat he is listening to. Plenty of humans cannot match his choreographic ingenuity.

Two particular features of Snowball's dancing caught Patel's attention for their scientific significance. First, whenever the music sped up or slowed down, Snowball would adjust his movements to stay in time. This is a classic test of synchronization because a dancer must listen for the timing between each beat and anticipate how soon he needs to make his move if he is to land at the same moment the percussion does. Even more impressive, Snowball could stay in time even if the beat dropped out entirely. In order to stay in time during breakdowns, dancers must continue hearing the rhythm in their minds so as to be on beat when the drums reenter. Snowball thus became the first non-*sapiens* creature to unequivocally demonstrate beat induction.

Entrainment is a passive skill. All it requires is that we move in sync with a simple, evenly timed signal, whether a dripping faucet, a car's turn signal, or a ticking clock. When the pulse stops, entrainment stops and so does the listener's movements. Beat induction is an active skill. It requires that our mind *create* its own tactus out of a complex and ambiguous auditory pattern and maintain this "mental" rhythm when the beats go silent during a breakdown. Snowball handles breakdowns without a hitch, as you can see at the 0:32 mark in his performance to "Everybody (Backstreet's Back)." He continues to bob in time with the groove *in his head*. His movements are still in perfect sync when the silence ends and the drums abruptly return.

Snowball humbled scientists with his inventive moves and downgraded our inflated sense of our own uniqueness. All we can say now is that humans are the only *primate* who can engage in beat induction. Perhaps surprisingly, monkeys and apes can't keep time to a beat. One chimpanzee and one bonobo have been documented performing entrainment, but their synchronization abilities were limited to a single preferred tempo. No chimp (so far) does what Snowball and a California sea lion named Ronan can do: keep steady time to complex rhythmic patterns across a range of tempos (including dynamic tempos)—and stay on beat during breakdowns.

4

So how do a little bird's subversive dance moves shed light on Mathieu's beat deafness? Patel and his collaborator John Iversen used Snowball's dancing prowess to form an influential theory of beat induction known as "action simulation for auditory prediction" (ASAP). ASAP suggests that our subjective perception of rhythm—our personal experience of tactus—is *not* the result of repurposing our brain's internal clock. Instead, tactus depends on a special neural loop linking our auditory system with our motor system. Why does this loop exist in the first place? Could humans and cockatoos have evolved special brain circuitry for pop-locking or line dancing? Patel and Iversen made note of a possible clue: the fact that our auditory and motor systems both contribute to social communication, including body language, speech, and singing.

In humans, a thick pathway connects the auditory system to the premotor cortex, the brain system that instigates our movements, including those of our lips, tongue, and larynx during vocalizations. A second, bidirectional pathway connects the auditory cortex to the inferior parietal lobule, a region involved in determining *what is this sound to me?* and engaging our muscles for a rapid response to threats. In other primates, these two-way paths are greatly diminished.

ASAP contends that your urge to move to a beat arises whenever you hear a stream of regularly spaced acoustic events. First, your auditory cortex notifies your premotor cortex that your auditory cortex hears a pulse. Next, your premotor cortex instructs your motor cortex to tap your fingers or nod your head to the beat. As you move rhythmically, your auditory cortex exchanges information with a region of your parietal lobe that predicts when the next beat will arrive. Once your parietal lobe predicts the timing, the auditory system's focus on the rhythm is enhanced. Your auditory cortex evaluates whether your tapping is synchronized with the arrival of each beat. If they're in sync, congratulations! You've got rhythm! If not, your auditory cortex recognizes the mismatch and alerts the motor cortex to either speed up or slow down your body's movements to match the perceived beat.

For humans—and Snowball, apparently—this reciprocal loop of neural activity between auditory cortex and motor cortex triggers reward circuits when the body's movements are synchronized with the groove. Dancers experience a surge of satisfaction from success-ful beat induction.

Thus, according to ASAP, during rhythm perception your brain constructs a tactus out of a sophisticated evaluation of a perceived *pattern* rather than matching an internal timekeeping pulse to an external musical pulse. Just as your visual brain constructs a three-dimensional representation of a visual scene from the two-dimensional patterns impinging on your retina, your auditory brain builds a complex personal tactus from the auditory patterns imping-ing upon your eardrums.

ASAP makes several predictions. First, simply *imagining* a beat should engage a person's auditory system and produce neural pulses at fixed intervals. And so it does. Participants in a brain scanner remained motionless while listening to a repeating sequence of two identical tone pips followed by a brief gap of silence. Then these listeners were asked to imagine that the series of tone pairs formed a musical beat. In half the trials, the participants were instructed to imagine an accent on every first pip. In the other trials, they were to imagine hearing an accent on every second pip. Their brain activity revealed significantly stronger neural pulses whenever an *imaginary* accent occurred. Beat perception varied, depending on the listen-ers' consciously determined tactus, even though the objective pat-tern of the tones was unchanged across the trials.

The second prediction follows from the first: beat induction is under voluntary control. We can deliberately extract a perceived beat from an ambiguous auditory input or imagine a beat in the absence of a percussive hit, such as experiencing the "missing" downbeat in a conventional "one drop" reggae rhythm. (You can hear an example on Bob Marley and the Wailers' "No Woman, No Cry.") In this beat, the hi-hat plays a busy pattern while the kick drum stays silent for the "one" and lands only on the "three": one-and-a-two-and-a-THREE-and-a-four-and-a. Our capacity to perceive regularity in the missing beat makes dancing to reggae easy and pleasurable.

Finally, ASAP makes a prediction that surprised many music-cognition researchers: we should find beat induction in animals who are competent at vocal learning—in creatures who can reproduce another creature's vocalization. Snowball was the first creature to vividly confirm this prediction, though scientists now recognize that psittacines (birds in the parrot family), pinnipeds (seals, walruses), cetaceans (whales, dolphins), and elephants are all vocal learners who possess neural circuitry supporting the perception of rhythm and the extraction of a tactus.

ASAP also illuminates Mathieu's beat deafness. A follow-up study with Mathieu recorded his brain activity while he listened to music. His brain did not exhibit any noticeable flaws in the early stages of processing music. He could *identify* a record's beats as well as anyone. Instead, his brain exhibited irregularities in higher-order processing—the same sort of "big picture" mental flaws found in other pattern-processing disorders, such as dyslexia, attention-deficit disorder, and tone deafness. These disorders have been labeled "perception without awareness." A person with these impairments can recognize individual items just fine (such as correctly identifying letters, pitches, or beats) but struggles to recognize sequential relationships that organize individual items into collective patterns (such as words, melodies, or rhythms). Thus, Mathieu's beat deafness is due to a faulty learning mechanism responsible for "attentional capture"—in this case, a rhythmic pattern-recognition circuit that must be intact to perform beat induction.

Furthermore, ASAP explains why Mathieu can successfully keep time with metronomic rhythms, like those found in dance pop and techno. His brain doesn't need to perform any sequential learning to extract a tactus from these genres' simple, unaccented, repetitive beats. Similarly, a person with tone deafness has no trouble recognizing that the pitch of E is different from the pitch of C. But ask him to identify a sequence of tones—for instance, whether the melody for "Row, Row, Row Your Boat" is the same as the melody for "Mary Had a Little Lamb"—and his tone-deaf brain may struggle to answer correctly. Although it might have been distressing for Mathieu to learn that he was beat-deaf, it must have been comforting to know that the reason he always seemed to be an "Elaine" on the dance floor wasn't due to any failure of willpower or lack of physical coordination.

He simply possessed a biological quirk that prevented his brain from organizing beats into a tactus.

5

It's New Year's Eve. You're at a lively party whirling with people, food, drink, and a DJ spinning the latest dance tracks. You might be the first one on the dance floor—or maybe you need to be dragged out there. Either way, the music is good, and you're enjoying moving your body along with the crowd.

Now let me ask you a personal question: What sort of dance are you doing?

Is your go-to move an up-and-down pogo-stick motion of the kind induced at rock concerts? Or do you prefer to lower your hips and involve your knees, like the dancers in Missy Elliott's "Lose Control (feat. Ciara & Fat Man Scoop)" video? Do you do a side-to-side motion like in Silk Sonic's video for "Smokin out the Window"? Or do you move from only the waist up, using your hands and arms or perhaps performing a classic head toss, brought on by records like

"Killing in the Name" by Rage Against the Machine? Or do you smoothly slide your feet across the floor like James Brown?

Dancing is the most common and pleasurable way for humans to express their love of rhythm. For many listeners, a record's danceability is its most important quality. It has been argued that some styles of music make practically no sense unless they are accompanied by dance, whether purpose-built dance music like salsa or the frenzied mosh pit of hardcore punk. The human affection for dancing has prompted many music scientists to declare that we cannot truly understand our relationship with music unless we consider how music makes us *move*.

Dance music usually features a 4/4 or 2/4 time signature. There's a biological explanation for this. The bilateral design of the human body is perfectly suited for moving in sync with an even number of beats per bar. Almost all human locomotion, including crawling, walking, jogging, and sprinting, alternates between moving the limbs on the left side of our body and the limbs on the right side. Walking, for example, employs four movements in a repeated pattern: right foot down, left foot up, left foot down, right foot up. If we count each up or down motion of a leg as a distinct beat, then we naturally walk to a 4/4 time signature. If your foot hits the floor on the downbeats—ONE-two-THREE-four—then your knees will reach their maximum height on the backbeats—one-TWO-three-FOUR.

Dance music relies on the fact that our bodies have certain rates at which we prefer to move our symmetric limbs. It was once thought that humans preferred musical tempos that match our heart rate, but that's not true. The adult resting heart rate hovers around an average of 72 beats per minute (bpm). Adults' preferred dance tempo, however, is closer to an average of 123 bpm, roughly the speed of a brisk walk. (If you would like to find your own sweet spot for tempo, visit our website, ThisIsWhatItSoundsLike.com, which contains a link to a testing tool. The average spontaneous tapping rate across listeners is around 100 bpm.)

The dance you instinctively feel like performing on the New Year's Eve dance floor is dictated in large part by your personal perception of a record's rhythmic accents. Accents are created by modulating the duration and loudness of percussive hits to produce distinctive strong and weak beats, making it easier to interpret a rhythmic pattern, such as STRONG-weak-STRONG-weak. Accents also make it easier for listeners to grasp larger rhythmic schemes, helping us predict when the next musical *section* will begin, not just the next beat.

Accents can cause musical rhythms to mimic speech. Languages differ according to whether they are stress-timed or syllable-timed. Stress-timed languages, like English, Russian, and Arabic, put accents on certain syllables, regardless of when they arrive in a sequence. Aniruddh Patel invites us to consider the accents in the English sentence "THE TEAcher is INterested in BUYing some BOOKS." There are an irregular number of unaccented syllables between each pair of accents—sometimes one, sometimes two, sometimes three, sometimes zero. In a stress-timed language, speakers will often jam all the unaccented syllables together so that the accents arrive at a steady pace: ONE-and-TWO-and-THREE-and-FOUR-and / THE-[and]-TEA-cheris-IN-terestedin-BUY-ingsome-BOOKS-[and].

In contrast, syllable-timed languages, such as French, Spanish, and Yoruba, place accents between consistently spaced syllables. The same sentence in French is "Le PROfesseur esT INTéressé À ACHeter des LIVres." There are exactly three unaccented syllables between each accented syllable. Thus, the tactus we naturally experience while listening to syllable-timed languages is more consistent than the tactus we experience from stress-timed languages. As we saw in the Melody chapter, this disparity likely accounts for the way the rhythm of a particular musical composition seems to reveal the native language of its composer.

Although they are sometimes obvious, musical accents on records can be subtle. Here are two dance tracks, both in 4/4 time, with different accent patterns. As you listen to each record, pay attention to where you hear the "weight" of the rhythm in each bar. First up is "Levitating (feat. DaBaby)" by Dua Lipa. You may hear the kick drum on the one and three as slightly longer than the claps on the two and the four: LONG–short–LONG–short. If you were dancing to this track, you might feel the urge to emphasize your movement on the accented downbeats, with the kick (ONE-two-THREE-four). In contrast, listen to "HandClap" by Fitz and the Tantrums. Here, the kick drum is a tad shorter than the snare, which we perceive as lengthened because the snare is doubled with hand claps: short–LONG–short–LONG. This record wants you to clap on the backbeats (one-TWO-three-FOUR), or on every beat when the soul claps pop up in the chorus.

Another rhythm-related variable that sorts listeners into different camps is *what part* of your body feels the urge to move. Metal and rock music lovers who are deep into the groove might use their arms to play air guitar or air drums, or at least do some headbanging. Funk music can bring on "pigeon neck"—where your head moves front to back like a pigeon pecking at seed. Samba music is aimed at our hips. The robot dance, with its "dimestops"—the abrupt halting of motion—imitates the mechanical action of machines, so it is expressed with the entire body. The hard-hitting intensity of hip-hop spawned full-body athletic moves, just as classical music spawned ballet, with its graceful—and just as athletic—leaps and arcs.

As new musical rhythms evolve, the way we dance evolves, too, as we discover new ways to feel the beat.

6

Music's rhythm rewards us by fulfilling—or violating—our expectations of when the next beat will arrive. One way records provide us with a reward-provoking surprise is by introducing a breakdown where the rhythm goes silent. Some styles of music use less conspicuous violations to spice up the record's rhythm. When a beat consistently falls outside a song's apparent rhythmic structure, this is known as *syncopation*.

Syncopation is like an optical illusion for rhythm. It exploits the design of our brain's beat-induction circuitry to trick us into perceiving beats that aren't actually there. A record featuring syncopation often has silence where we expect beats and beats where we expect silence.

Imagine that a walking motion (right foot down, left foot up, left foot down, right foot up) represents a straight (non-syncopated) accented rhythm of ONE–and–two–and–THREE–and–four–and. The "and"s represent the motion *in between* the main beats—when our limbs are moving into the next position. A syncopated rhythm emphasizes some of the transitional moments rather than the culminating moments, such as: one-and-two-AND-three-and-four-AND. The strongly accented "AND"s form the syncopated beats.

A syncopated rhythm might feature the kick or snare putting its beat on the "AND"s and omitting the usual ONE or TWO. Our beat-induction circuitry knows that a beat *should* land in those moments where we perceive silences, so our brain allows these omissions to serve as beats. If this pattern of missing-yet-perceived beats repeats in every bar, we can easily dance to the syncopated rhythm.

Syncopation is often called a rhythm within a rhythm because our bodies naturally want to move to downbeats and upbeats rather than in-between beats, even when the downbeats and upbeats are not actually performed. The acclaimed scientist Tecumseh Fitch asserts that syncopated rhythms "inject energy" into the moments when we transition our arms or legs to the next event, causing our corresponding dance movements to accelerate and decelerate in more complex ways.

Nearly every known style of music features some degree of syncopation, though some styles—such as Latin and African music— exemplify it. You can hear reggaeton syncopation in the drum track of "Yo Perreo Sola" by Bad Bunny. Listen to this track and pay attention to how your body chooses to move to it. Compare your tactus for this song with the tactus you felt when you listened to "Stoned and Starving." You may recognize that these two rhythms seem to direct your body into different kinds of motion: side to side during "Yo Perreo Sola" and up and down during "Stoned and Starving."

You can hear jazz syncopation in the skipping tap of the snare drum and the strongly accented hi-hat in "Poinciana (Live at the Pershing, Chicago, 1958)" as performed by the Ahmad Jamal Trio. You hear it in R&B with the hurry and pause of the bass in "Stay Flo" by Solange. Rock music doesn't often feature syncopation, but you can feel syncopated beats in the classic punk song "Lust for Life" by Iggy Pop. Rock and punk music typically elicit a pogo-stick motion, but in the official "Lust for Life" video, dancers bend their knees, twist their feet, and rotate their hips, their movements more closely resembling those of a traditional West African dance than pogoing.

Syncopation requires your brain to do more work during beat induction, but for many listeners this extra effort is all the more entrancing to their rhythmic sweet spot.

7

So far in this book, we've focused on your *personal* experience of music listening. But when it comes to enjoying music with a crowd, rhythm always gets the party started. Ethnomusicologists tell us that our prehistoric ancestors took advantage of the natural allure of rhythm to build tribal unity. Clapping two rocks together in time (*concussive* noise) or thumping out a steady beat on a hollow log (*percussive* noise) invites a community of people to move in synchrony.

A group of people moving as one sends the powerful message that everyone is feeling the same emotions, just as a group of people singing the same lyrics suggests they are thinking the same thoughts.

For hundreds of thousands of years, music making helped families establish bonds, especially mother-infant bonding through lullabies. Eventually, music may have contributed to tighter social connections by encouraging early humans to live in communities rather than roaming alone. The powerful magnetic draw of synchronized choreography may explain why group dances are a staple of every culture. Much of the pleasure derived from watching Irish folk dance, Texas line dance, Palestinian *dabke*, Polish polonaise, Indian *bhangra*, or the synchronized steps of a boy band comes from identifying with a community and experiencing a sense of belonging.

Anthropologists have found that people laboring as a group often fall into a shared physical rhythm, moving their bodies in synchrony as they work. Sometimes these crews add chants or other vocalizations. Adorning our rhythms with words and melodies can further heighten the sense of community. When we make music as a group, we tend to move in unison, form similar facial expressions, and even breathe together between lyrical phrases. Communal music making bypasses the need to express our musical selves as individuals, letting us fuse our identities with something larger than ourselves.

Of all the ways in which we emotionally bond to music, rhythm perception is the most fundamental. A record doesn't require a heart-wrenching melody, a take-your-breath-away lyric, or a chill-inducing timbral design to knock you out and earn your devotion. A groove that matches one of your personal sweet spots for rhythm can make a record feel as though it were tailor-made for your body. When a beat fits you like a bespoke suit, you've found true musical love.

Musical Development

Music can help children acquire language, develop physical coordination, and learn social norms. Just as children's bodies grow more and more coordinated as they move through and interact with the physical world, children's auditory systems become increasingly efficient through exposure to the complex sound patterns of music.

Our auditory system reaches its peak perceptual ability in our late teens. Rapid growth happens between the ages of eight and eleven, as the brain prepares itself for the upcoming changes that puberty will introduce. Without training, the musical perceptual aptitude of most people plateaus before their teens. Studies show that the performance of eleven-year-olds on music-perception tasks is the same as that of adults who were never trained in music.

Getting children involved with music early isn't merely a tactic for boosting their odds of getting into a good college; it can also lead to greater emotional and perceptual sensitivity. Children who take music lessons exhibit more prosocial behaviors, like sharing, cooperating, and empathy, than children involved in other activities. They are also better able to learn by listening.

CHAPTER 7

TIMBRE

This Is What Music Conjures

✧

Her voice was ever soft,
Gentle, and low—an excellent thing in woman.

—*William Shakespeare*, King Lear

1

THE NAME STRADIVARIUS IS PRACTICALLY SYNONYMOUS with the world's finest violins. But have you ever wondered what makes a Stradivarius so exorbitantly priced compared to the latest mass-produced violin at the local music store? For that matter, why would someone pay more than the cost of a Tesla for an original 1959 Gibson Les Paul Standard when a decent brand-new electric guitar can be had for one-hundredth the price? Valuable instruments earn their price tags due to several factors, including their rarity and the fame of the musicians who once played them. But the main reason that Stradivarii, vintage Gibsons, and other highly coveted instruments are so pricey is because one crucial quality sets them apart: their *timbre*.

Pronounced "TAM-ber," it is the most enigmatic dimension of music. Timbre refers to the unique *voice* of an instrument: the acoustic qualities that let us distinguish a guitar from a trombone, and a trombone from a tuba. Timbre is the sonic equivalent of the taste of fine wine (spicy with complex fruit flavors) or the scent of perfume (floral with patchouli notes). And just like your response to the flavor of a bottle of Romanée-Conti or the aroma of Flowerbomb by Viktor&Rolf, your response to a particular timbre is unique to you. The perceived quality of a timbre ultimately comes down to personal taste, prior experience, and, on occasion, social influences—such as its price tag.

The violin as we know it today was invented in the early sixteenth century in the mercantile city of Cremona, Italy, by Andrea Amati (1505–1577). He set the bar for timbre and playability, attracting many apprentices and followers. These pioneering luthiers (makers of stringed instruments) consciously strove to build instruments that could "rival the most perfect human voice." During his lifetime, *maestro di Cremona* Bartolomeo Giuseppe Guarneri (1698–1744) was considered to have come the closest to perfection, earning him the nickname "del Gesù" ("of Jesus") for the sonorous baritone timbre his instruments produced.

For his own part, Antonio Stradivari (1644–1737) took up the steep challenge of improving upon the Amati violin. He lengthened the signature *f*-hole and experimented with new varnishes to craft an instrument whose timbre was comparable to that of a human singer. Compared to the rich baritone timbre heard in Guarneri's violins, Stradivari's instruments produced a higher resonant frequency that more closely resembled tenor and alto singers. This human vocal quality was a major reason that musicians and audiences in Stradivari's day—and for centuries afterward—came to regard his violins as possessing an intangible acoustic character that elevated them above all the rest.

Some readers might be wondering, *Okay . . . so why don't modern luthiers just follow the same recipe as Stradivari and churn out more Stradivarius violins?* A reasonable question. The answer: it's impossible.

Violins made by the Stradivari, Amati, and Guarneri families were built from trees cut down in Renaissance Italy and varnished with materials made from that era's local plants. The biosphere during the golden age of violin making was markedly different from what it is today. Biochemical compounds in the water, soil, and air were untainted by soot, sulfur dioxide, carbon monoxide, heavy metals, plastics, and petroleum. The presence of these pollutants causes plant cells to develop differently and to produce a different sound quality when converted into instruments. Centuries of smog and acid rain have made it unfeasible to obtain the same kind of wood and varnish used by Stradivari.

But are the timbres produced by golden age violins truly superior to those of modern instruments . . . or does the name and sticker price make us *think* they are? Claudia Fritz and her colleagues at the Université Paris Cité were brave enough to tackle this question in 2012.

Twenty-one expert violinists in town for an international competition were asked to judge six violins. They evaluated three new top-of-the-line models, each from a different maker, and three classic ones: one Guarneri del Gesù (ca. 1740) and two Stradivarii (ca. 1700 and ca. 1715). The older violins had a combined value of $10 million—roughly one hundred times that of the new violins. The instruments' identities were kept secret by auditioning them in a darkened room and having the players wear modified welder's goggles. The researchers even went so far as to dab a bit of scented oil on every chin rest to mask any telltale odors. After playing the violins for as long as they wished, seventeen of the violinists were asked to report their preference: *Which instrument did you like best?*

For playability, projection, and response, there was a clear answer. Violinists preferred the twenty-first-century models. That was not a major shock, as one can easily imagine how a newer violin might offer greater comfort and control. But the big question remained: Which violin's *timbre* did the performers like best? This time, the researchers were in for a surprise.

The Stradivarii were *not* ranked at the top of the list. Instead, all six instruments exhibited significant overlap across the musicians' judgments of the richness of tone—though the lowest timbre score was assigned to a Stradivarius.

Could these elite violinists guess the era of their preferred violin? Remarkably, when asked whether the violin they'd most like to take home was old or new, only three of the seventeen respondents correctly identified the instrument's era. Seven had no idea, and seven guessed incorrectly (mistaking an old violin for a new one, and vice versa). In short, trained experts couldn't distinguish between the sound of a freshly minted digital age instrument and a handcrafted instrument that was three hundred years old.

How could it be that the timbres of the renowned Cremona violins weren't preferred over that of the modern instruments? Given the stir created in the popular and scientific press that resulted from their paper, the researchers were compelled to respond. They noted that the violinists' judgments of timbral quality were made "under the ear," meaning that the players rated the instruments based on how they sounded as they were playing them. The experiment didn't provide any insight into how the different violins might have sounded during a live performance in a concert hall, under the same conditions that an audience would experience.

To see if their findings would hold, Fritz's team followed up with a new study of similar design but with greater experimental validity. This time they used twelve violins—six new and six old (including five by Stradivari)—and let ten expert soloists put them through their paces in both a rehearsal room and a concert hall, under blind-test conditions. After playing the instruments for over an hour, each soloist was asked which of the twelve they would choose to take on a concert tour. Six of the ten performers chose a new violin. The most consistent praise was heaped on a twenty-first-century instrument. Mostly what was preferred about the new violins was their ease of use, but as in the first experiment, there was no significant preference for the timbre of one violin over the others. Once again, the experts could not distinguish the timbres of new violins from those of antique violins.

In fairness to Stradivari's reputation, it is estimated that only five hundred of his violins are still in existence, and most of these have been repaired or modified. We cannot say with certainty that the timbre they produce today is the same as what Antonio Stradivari heard when he built them. These studies do nothing to reduce the historical importance or cultural value of golden age violins.

Nevertheless, their unforeseen findings are a reminder that when it comes to judging the quality of timbre—deciding whether we *like* a particular sound—a large portion of our judgment will always be influenced by biases and preconceptions. It's the very fact that our brains can identify such a vast diversity of sounds—the screech of old brake pads, the babble of a forest stream, the electronic chirrup of an old-fashioned dial-up modem—that makes timbre *perception* so complicated for researchers to model and so difficult for listeners to agree on.

Timbre is a *continuous* (or "analog") perceptual phenomenon: timbres can be manipulated to cross perceptual boundaries and become new sounds, such as modifying an acoustic guitar tone to make it sound like an electric guitar. But in the transition from an acoustic timbre to an electric timbre, the guitar sound proceeds through a highly nuanced spectrum of in-between timbres, each producing its own mental effects on the listener. This contrasts with melody, lyrics, and rhythm, which are *discrete* (or "digital") perceptual phenomena. Western tonal music recognizes twelve distinct pitches in an octave. The frequencies midway between each standard pitch (called *microtones*) are rarely used. Nearly every word in a lyrical verse is distinct, except in certain cases of alliteration. A rhythm is comprised of discrete beats. The very notion of a "continuous" beat is nonsensical. But tweak any timbre, even slightly, and it becomes a whole new sound, one that can be usefully incorporated into a record.

We can notate melodies, lyrics, and rhythms of songs in scores or demonstrate their exact values for someone by singing the pitches, words, or beats. But how can you describe a timbre to someone else so that she can reproduce its sound exactly? A guitarist once asked me to put more "r r r r r" on his guitar tone. Another suggested that a mix was "a little orange" and he wanted it to be "bluer." Prince would ask for more "sauce" on a sound. These imprecise requests for a specific timbre reflect its confounding nature and how hard it can be to characterize a timbre in words.

The difficulty of notating timbre and its infinite range of acoustic qualities is why the seventh and final dimension of your listener profile has a greater impact on *records* than on songs. When it comes to timbre, you just have to hear it.

2

Timbre is a potent catalyst for memories. Hearing a rusty squeak can instantly take us back to childhood and, say, the memory of a creaky screen door. The soft gurgle of oars in a lake may remind us of paddling a canoe at summer camp. The distinctive memories we associate with any given sound is why changing a record's timbre can change its impact on the listener, even though the song—its melody, chord progression, time signature, and lyrics—remains the same. A good example is "Hurt," a 1995 recording composed and performed by Nine Inch Nails.

The record features Trent Reznor's delicate, breathy voice describing how he hurt himself just to feel something deeply. Though the song's meaning is disputed by fans, it is generally considered to be about self-harm, likely related to drug addiction. We hear a young man practically whispering about his pain, as if telling us a secret. His shallow inhalations are audible on every line. He *sounds* like someone who is hurting.

Johnny Cash released his own version of "Hurt" in 2002, showcasing a very different vocal timbre. The deep and sonorous resonance of Cash's voice is a stark contrast to Reznor's thinner timbre. Cash's vocals rasp like a locomotive lumbering up a hill, bringing to mind a man with the weight of life and history bearing down on his aging body. When my students listened to the two versions in class, most reported that they found Cash's performance more compelling and chilling than Reznor's. The timbre of Cash's voice evokes an older man who is sure of himself and means exactly what he says, compared to a floundering younger man who may be lost and is pleading for help. When Cash utters the line "I remember everything," the aching tremble in his voice expresses a lifetime's worth of eventful experiences. Cash's mature, weathered timbre elevates the stakes by enriching the subtext behind the performance, making the record more potent.

Although Reznor had written the song as a deeply personal statement, Cash's version shocked him. Reznor compared the feeling of hearing Cash sing his lyrics to that of losing a girlfriend to a rival. "That song isn't mine anymore," he conceded.

Modifying a timbre can redistribute a song's emotional weight. That is why the "unplugged" (acoustic) version of a well-known record can be so pleasing, provided the song is a good one. The beautiful melody of "Teardrop" by Massive Attack is wrapped in drum machine, harpsichord, piano, and record scratches to create a dreamy soundscape as a backdrop to the ethereal timbre of Elizabeth Fraser's voice. José González recorded a version of "Teardrop" with very different timbres. By simplifying the accompaniment and redesigning the timbral palette around acoustic guitar and his stronger voice, González's record highlights the mesmerizing melody.

Timbres and their arrangement influence our perception of the relative importance of a song's melodic, lyrical, and rhythmic elements, mediating the rewards we derive from the other musical dimensions. In 1984, I heard this influence firsthand in dramatic fashion. When Prince first recorded "When Doves Cry," it was as thickly layered as "Darling Nikki," an explosive pop/rock track also from the album *Purple Rain*. "When Doves Cry" entered the world with a full menu of high-intensity timbres, including distorted keyboards and distorted guitar. The lead vocal was the lightest timbre in the mix. Prince initially chose to emphasize heavy sounds to summon the power of rock music and a muscular rhythm.

However, if you hum the song's chorus—"How can you just leave me standing / Alone in a world that's so cold . . ." —you'll hear that its melody tilts toward the rhythmic and away from the melodic character heard in most "singable" choruses. Prince recognized that without lighter timbres, the lead vocal would simply get lost. The original arrangement of the song was too bombastic to match the gentle and sorrowful emotions expressed in its lyrics. Distortion contradicted the main message of the song: *This is what it sounds like when doves cry.*

Prince revised his approach, starting with the timbres. The distorted rhythm guitar and keyboards were stripped away. In a bold and ingenious move, the bass guitar was muted, too. He realized that all the heavy timbres could be culled without losing intensity or meaning. Lighter timbres were a better fit for the spirit of the lyrics and allowed the drum pattern—kick, side stick, snare, claps, flanged hi-hat—to dominate the record. The version that we all know became Prince's first No. 1 single on the *Billboard* Hot 100, and his attentiveness to timbre was a big reason why.

Given timbre's ability to evoke personal associations, it's little wonder that hearing certain musical instruments can prompt memories of previous records that promoted the same instrumental timbre. Sleigh bells can call to mind the Beach Boys' "God Only Knows." Glockenspiels can evoke Bruce Springsteen's "Born to Run." Kalimbas may make you recall Gotye's "Somebody That I Used to Know." Depending on your age, a vocoder might conjure up Kraftwerk's "The Robots," the Beastie Boys' "Intergalactic," or Imogen Heap's "Hide and Seek."

Because your personal associations with timbres are different from mine, we each judge the quality of a timbre differently. What sounds dated and old-fashioned to one listener can sound fresh and bold to another. For example, the Moog (pronounced "Mohg," not "Mooog") synthesizer was practically synonymous with progressive rock in the 1970s. You can hear its iconic timbre in the record that originally brought it to prominence: *Switched-On Bach* by Walter (later Wendy) Carlos. Moogs tumbled out of favor in the '90s as a new generation of teens preferred grungy guitars to the synthesized keyboard sounds favored by their parents. But the Moog sound came roaring back in the 2000s as the Moog Music company regained its financial footing. Artists as diverse as Radiohead, Alicia Keys, Muse, and Stereolab eagerly added new Minimoogs and Moogerfoogers to their gear collections and vintage Moog synthesizers fetched record-setting sums on eBay.

In contrast, the kick drum sample in the Roland TR-808, a cheap drum machine introduced in 1980, has a popularity that is apparently unquenchable. It achieved iconic status after Afrika Bambaataa and Soulsonic Force's hit single "Planet Rock," pushed hip-hop onto the pop charts. Roland ceased production of the TR-808 in 1982, after a mere two years, but, ironically, that may have contributed to its cultural success. As more sophisticated drum machines came along, musicians unloaded TR-808s at secondhand music stores, putting them in the price range of young, home-based record makers. Its ubiquity in hip-hop, rap, and electronic music was still driving people to the dance floor in 2003 with Outkast's "The Way You Move," which even featured the call-out lyrics "But I know y'all wanted that 808!" You will surely recognize it in "God's Plan" by Drake, released in 2018. The satisfyingly pure tone of the 808 kick drum is a timbre that is somehow like no other, and has become a sonic meme that spans generations.

We associate many talented musicians with their unique sonic signatures. A familiar timbre may even set expectations for the performance. For fans of classic rock, a distorted Gibson Les Paul guitar immediately brings to mind Keith Richards of the Rolling Stones or Jimmy Page of Led Zeppelin. Blues guitarist B. B. King popularized the mellower timbre of the Gibson ES-335 on the instrument he named "Lucille," and so older listeners may associate that tone with the blues. The Fender Stratocaster's mournful cry, sublimely expressed by Eric Clapton and his beloved Strat "Brownie" on "Bell Bottom Blues," is a staple of later blues rock music. Country music pioneer Chet Atkins popularized the sound of the Gretsch Country

Gentleman, a large, hollow-bodied electric guitar immediately recognizable for its warm, round twang. James Hetfield, a guitarist with Metallica, helped launch the sound of metal with the buzz-saw timbre produced by his ESP Snakebyte fed through a Mesa Boogie amp. Musical timbres trickle down to influence every young musician whiling away a Saturday morning in a music store.

Because our associations with sounds are so personal and evocative, it's little wonder that musical timbres can have a powerful emotional effect on us. Life teaches us that certain sounds are likely to be heard only in certain contexts, so when a sound is unexpected or out of place, it can trigger fear. The most obvious fear-evoking timbre is a human scream.

Screaming occupies a "privileged niche" in our mental architecture, instantly firing up our danger-detection circuits to alert us to an urgent threat. Because screaming is usually heard in dire contexts, your nervous system has learned to drop whatever it's doing and orient toward this singularly arousing timbre. Music makers are wise to our predisposition and will often emphasize a climactic emotional moment on a record using a sound that mimics screaming. Listen to the guitar solo on Radiohead's "Paranoid Android" starting at the 3:04 mark. When Thom Yorke's vocals leave off and guitarist Jonny Greenwood leaps into the fray with his piercing electric guitar, he creates a moment that we unconsciously associate with an *in extremis* situation.

Famous movie scenes can get linked in our imagination to the musical timbres that accompanied them. If a scene is dramatic enough and the sound distinctive enough, the timbre can even assume the authority of a cultural norm. Consider the famous shower scene from the classic horror film *Psycho*. A pulsing shriek of violin bows violently dragged across strings plays as the actress Janet Leigh is attacked by a psychotic Norman Bates. The hair-raising, scream-like sound influenced every horror film composer who followed in the wake of *Psycho* composer Bernard Herrmann.

Early film composers used the harp to underscore heavenly scenes. Today its fluttery glissando has come to parody the ecstasy of love or the swoon of an altered mental state. Perhaps due to its size or its extremely low range, the tuba has been associated with comedy since the 1950s. It was delightful to encounter a tuba in the recording studio while working with David Byrne on his song "My Love Is You"—a droll yet sincere expression of how the beloved's imperfections are what make her perfect for the singer. David called in acclaimed New York tuba player Marcus Rojas to add brass to the arrangement. Before this session I hadn't known that, in the hands of a professional like Marcus, the tuba could laugh, squeal, and play the role of sidekick in David's lyrical story. Working together, player and instrument effectively communicate with listeners by welding a musical idea to its ideal timbre.

Timbral associations are so strong that a music maker must be cautious when using an unusual timbre, in case listeners may inadvertently tap into preexisting associations. Los Angeles record producer Tony Berg owns the finest collection of musical instruments I have ever seen. Among his treasures is a bass harmonica, a classic instrument made famous on the Beach Boys' 1966 album *Pet Sounds* (you can hear it during the intro of "I Know There's an Answer"). Unfortunately for Tony and others who love this instrument, it was also made famous on the 1960s television sitcom *Green Acres* as the musical accompaniment for the pig named Arnold Ziffel. I dubbed Tony's harmonica "Arnold" and couldn't resist mentioning its namesake pig every time we recorded it.

3

Why are our brains so adept at identifying timbres? One school of thought suggests it is because our mammalian ancestors were once entirely nocturnal and had to rely on sound rather than vision to identify objects in the dark. Day-living primates eventually emerged, leading our brains to develop exceptional vision, too, but our visual capabilities appear to have been stacked on top of even more ancient audio capabilities. We were good at listening before we got good at seeing.

To fully appreciate how we perceive timbre and why it forms such a highly influential dimension of our listener profile, let's trace out how your brain processes simultaneous sounds. One of my science heroes, Professor Emeritus Albert Bregman of McGill University, literally wrote the book on auditory scene analysis. His classic monograph, appropriately titled *Auditory Scene Analysis*, explains how listeners pick out and follow individual instruments, voices, and environmental noises in the ever-changing bricolage of sound streaming into our ears.

Consider the mental challenge that your auditory system must confront when trying to isolate and identify sounds in a crowded soundscape. Imagine, for instance, that you are reading this book in the middle of a busy outdoor café. Though your brain is focused on the text in front of you, you hear a wide variety of sounds coming from every side: people chatting, mugs clinking, footsteps tapping, keyboards clacking, cats hissing, horns honking, and traffic rumbling. Your brain knows from prior experience that these noises aren't unusual for an outdoor café, so it's easy to tune them out and focus on your book.

Suddenly, a musician starts playing a melody you love on the piano. In an instant, your attention goes straight to the music. You ignore all the other ambient sounds as your mind prepares to savor the sweet chord change you know is coming. How were you able to effortlessly isolate the piano tones from the omnidirectional cacophony of noises, even if you weren't expecting to hear music?

Singling out a particular timbre from a complex web of sounds is a very different sort of mental challenge than choosing where to focus

Auditory Scene Analysis

your vision. Objects in your visual field occupy distinct locations. When light reflects off various objects and surfaces, the rebounding photons activate different segments of your retina. Your brain can use these topologically distinct visual boundaries to map out the visual scene. In contrast, auditory objects—clinking utensils, chattering voices, and a melodious piano—reach your eardrums packed into a single composite soundwave. If vision were like audio, you'd see images of cats, cars, shoes, mouths, laptops, and a piano all juxtaposed over one another, like a stack of slides stuffed into a single slot in a slide projector.

All of the properties of sounds you can consciously focus on—melody, words, rhythm, timbre, loudness, spatial location, movement—are extracted from just three types of information embedded within the soundwave: *frequency*, *amplitude*, and *phase*. Phase refers to the relationship between the soundwaves arriving at your left ear and right ear. It's primarily used to localize the source of a sound and determine which way it is moving. All the other subjective properties of a soundwave (including the four musical dimensions of melody, lyrics, rhythm, and timbre) are derived from its frequencies and their moment-to-moment amplitudes.

Frequency is a measure of how fast a soundwave is vibrating at the instant it excites your eardrum. Amplitude is a measure of the soundwave's intensity. These two simple properties completely dictate our experience of music. Our brain can distinguish every Gregorian chant, every Wagner opera, every Charlie Parker solo, every thrashing punk anthem, and every Bollywood love song from every other based solely on the pattern of amplitude and frequency changes in their soundwaves.

Distinguishing between these different pieces of music is a complicated affair, however.

First, a soundwave vibrates your eardrums. Next, the vibrations of your eardrums are transmitted mechanically to your *cochlea*, the organ of acoustic perception, via the three smallest bones in your body. Next, a complex signal representing the wave's frequencies and amplitude ascends from the cochlea through the brain stem to reach the *primary auditory cortex*, which initiates an analysis of the audio scene.

The primary auditory cortex scrutinizes the pattern of frequencies and amplitudes in the arriving soundwave. With its dense connections to the motor, memory, language, decision-making, emotion, and reward systems of the brain, the auditory cortex next splits the sonic information into distinct auditory *streams* to send onward for further processing.

If you are sitting in an outdoor café, your primary auditory cortex might divide up the waveform arriving from your eardrums into a "conversation at the next-table" stream, a "footsteps on the pavement" stream, a "city traffic" stream, and a "piano music" stream. "The stream plays the same role in auditory mental experience as the object does in visual," observes Bregman.

Next, your brain must choose which stream to focus on. Some of

The Musical Mind

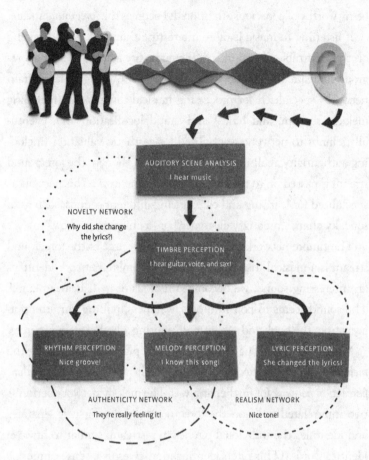

the acoustic sources at the café, such as the traffic, espresso machines, air conditioners, and pedestrians, create chaotic, unstructured sound patterns. Your brain has learned to ignore garden-variety ambient noises. But if a stream contains an orderly pattern of frequencies—known as a harmonic series, characteristic of pitched instruments and spoken vowels—then your brain takes notice. Your primary auditory cortex highlights the structured sound stream as a candidate for more attentive processing. If no other stream is making a stronger bid for attention (say, the intensifying rumble of a truck heading right toward you), then the structured stream gets selected as potentially being worthy of conscious attention. It becomes the *foreground stream*.

If listening to music is more interesting than, say, eavesdropping on the next table, the sounds of the piano are appointed as the foreground stream. The piano stream is then sent to higher-level brain networks specialized for processing musical dimensions (including melody, rhythm, and lyrics) and sound localization and, eventually, climbs to networks specialized for aesthetic valuation (including authenticity, realism, and novelty). First, though, the foreground stream is passed on to the *timbre-perception network*. This network is specialized for learning and categorizing different sounds, such as "a squeaky chair," "an angry customer," or "a chirping sparrow."

The timbre network recognizes that the source of the foreground stream is a musical instrument. Once the timbre network identifies an interesting sound, we become conscious of it: "I hear a piano!" The sound seems to pop into our awareness, pulling our attention away from the book and subliminally muting all other sound sources.

Next, the timbre network passes the piano stream on to the *melody-perception network*, *lyrics-perception network*, and *rhythm-perception network* for further analysis. The melody network performs two interrelated tasks on the foreground stream. First, it separates and identifies the individual pitches in a stream, similar to how we identify words. (This pitch identification gives us a sense of music's key.) Next, the melody network links the sequence of pitches into a single unitary melody, similar to how we group a string of words into a sentence. Finally, the melody network accesses its memory circuits to see if it has encountered this particular melody before. Can you name this tune? "I'm hearing the Brazilian classic 'The Girl from Ipanema.'"

The rhythm-perception network operates in parallel to the melody network. While the melody network attempts to identify the foreground stream's tune, the rhythm network performs beat induction on the same stream, creating your personal tactus. The rhythm network begins to make predictions about when the next accented tones will arrive. When the network successfully extracts a beat hierarchy from the piano stream, we experience a conscious sense of groove: "It's played with a syncopated beat."

If the pianist sings along with the melody he is playing, then your lyrics-processing network goes to work at the same time as your melody and rhythm networks. The lyrics network, which encompasses all the language-processing circuitry of your brain, operates similarly to the melody network. First, the lyrics network recognizes the individual phonemes sung by the pianist, then combines these phonemes into words, then combines these words into a single unitary verse, analogous to a melodic phrase created by the melody network. Next, the lyrics network attempts to comprehend the verse. If it's successful, we become conscious of the lyric's meaning: "He's singing, 'And when she passes, / Each one she passes, goes "Aaah." ' "

Finally, your brain integrates the outputs of the four music dimension networks to form a holistic representation of the music that integrates together the four musical dimensions of timbre, melody, rhythm, and lyrics: "I'm listening to a young male tenor sing 'The Girl from Ipanema' as he plays a well-tuned grand piano in a bossa nova rhythm."

The holistic music representation is then passed on to three higher-order brain systems that judge the music according to our aesthetic sensibilities. Each aesthetic system—authenticity, realism, novelty—encompasses multiple neural structures interconnected across the brain. Each aesthetic system receives holistic representations from all sensory modalities, not just streams of sound emerging from the auditory brain. Each system boasts its own connections to your reward circuitry. As a result, each aesthetic system can independently generate an experience of pleasure—or displeasure. Likewise, each of the four musical networks connect to your brain's valuation circuitry, letting you experience distinct rewards or disappointments from a song's timbre, melody, rhythm, or lyrics.

It takes less than 150 milliseconds for your brain to process an incoming soundwave, separate out a foreground stream, identify the stream's timbre, melody, lyrics, and rhythm, and unite these perceptions into a holistic conscious experience. Now that you are aware of the music, you can decide if you want to lean back and enjoy it for a while or return your attention to your book.

Learning to distinguish the dimensions of melody, lyrics, rhythm, and timbre begins in infancy.

Infants bang on pots and pans, drop things, shake their toys, shout, cry, whimper, and pull the dog's tail to learn about sound. When we're very young, we learn that different objects make different noises, depending on what they're made of and how we caused them to react. Hollow objects sound different from solid objects when you bang on them. Dropping a plastic cup on the linoleum floor sounds different from when you drop a glass. The plastic cup bounces around for a few seconds in a rhythmic pattern, but when we drop a glass, there is a single loud crash. Thus, as soon as we are born, our musical brain begins building "timbre templates" for how things sound. These early templates influence the development of the sweet spots on our listener profile.

As we are exposed to musical instruments, we learn how their timbres are related to each instrument's materials, shape, and volume, as well as the nature of the force that produces the instrument's sound. For example, when the padded hammer of a piano strikes the strings, the resulting timbre has a fairly loud attack (the initial energy of the acoustic wave). Once the hammer returns to its resting position, the tone loses energy quickly. We pluck a string on an acoustic guitar and the wave vibrates down the length of the string, as well as to and from the instrument's body. We learn that a nylon-string classical guitar produces a sound that is round and soft, while the steel-string guitar produces a bright, metallic sound. A bow dragged across the strings of a cello makes a warm and rough

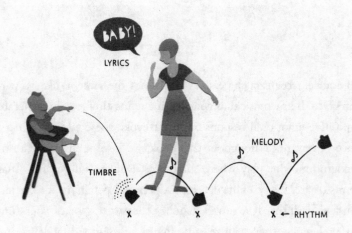

The Development of the Musical Mind

sound that will die away unless the cellist continues bowing to force energy back into the strings. The saxophone and oboe vibrate similarly to the human voice. Their unique timbres suggest that these instruments are played through a mouthpiece bisected by a reed, just as our somewhat reedy vocal cords bisect our "voice box," or larynx.

Timbre recognition is similar to face recognition. People are very good at matching pictures of adult faces with the correct face of the same person as a child. This is because our face-recognition circuitry recognizes the *relationships* between different visual elements in a face, such as the distance between the eyes, the shape of the nose, the presence and location of dimples, and the subtle asymmetry between the left and right halves of the face. Timbre recognition also relies on a global evaluation of all the relationships between the tonal elements of a sound. When we hear a singer for the first time, we build a one-of-a-kind timbre template that captures the unique pattern of harmonics in the voice.

That's why timbre is the "face" of music: it allows us to discern a sound's true identity.

4

Of course, *recognizing* a person's face is not the same as *liking* a person's face. If any musical dimension, including timbre, is to generate a positive emotional response, it must evoke the conscious experience of reward. Throughout this book, we have seen how pattern recognition is at play when our brains evaluate music and other forms of art. Every brain develops a personal penchant for certain musical patterns. If a sound matches the sweet spot of your timbre preference, then your reward circuits will respond by delivering the feel-good treat of dopamine release. The reciprocal connection between our timbre-processing network and our reward circuitry—and its vulnerability to external interference—is illustrated by the story of Mr. B.

Mr. B was a fifty-nine-year-old European with an unremarkable listener profile. His musical tastes were broad yet stable. His go-to records were those that had brought him the most joy in his teens, particularly music in his native language of Dutch. Mr. B also enjoyed the Beatles and the Rolling Stones, with a slight preference for the latter.

For forty-six years, Mr. B had suffered from a debilitating obsessive-compulsive disorder that made his life miserable. He was racked with anxiety whenever he faced uncertain outcomes or illogical events. He was compulsive about seeking a sense of reassuring control over his life, going so far as to hoard possessions. Given that his fears were disrupting his ability to function, his physicians determined that he would benefit from deep-brain stimulation of the nucleus accumbens, part of the brain's reward system—responsible for delivering dopamine bursts when we listen to music we like.

Neurosurgeons implanted two four-contact electrodes into Mr. B's nucleus accumbens. The effects were dramatic. For the first time in years, he was not seized by panic; nor did he feel the urge to behave compulsively. He began calling himself "Mr. B II" because of the new calmness, confidence, and assertiveness he felt. But these benefits were accompanied by a peculiar side effect. For the first time in his life, he became a devoted fan of Johnny Cash.

Shortly after his surgery, Mr. B was listening to the radio when "Ring of Fire" came on. His response was profound. The Johnny Cash record moved him more than music ever had. Mr. B attributed the powerful feelings he experienced to Cash's "raw and low-pitched voice."

Mr. B began buying and listening to every Johnny Cash record he could find. He noticed that the performances from the 1970s and '80s had the timbral quality that generated the most intense listening pleasure. No other music could satisfy him now: he reported to his doctors that "there is a Johnny Cash song for every emotion and every situation."

Unlike the intensely negative feelings stirred up by his pre-surgery obsessive disorder, Mr. B felt no manic obsession to listen to Cash, nor any anxiety during those times when he couldn't. Listening was *pleasurable*—not compulsive. Despite his exclusive and unremitting diet of Johnny Cash records, Mr. B claimed that Cash's music never became boring . . . until his implants ran down. When the batteries powering the electric stimulation of his nucleus accumbens died out, so did his passion for Johnny Cash. Once his batteries were recharged, so was Mr. B's interest in the Man in Black.

The strange case of Mr. B suggests that a sweet spot for timbre is intimately connected to our ability to experience pleasure. Our tightly knitted timbre-reward connection is also highlighted in the phenomenon of autonomous sensory meridian response, more commonly referred to as ASMR. ASMR has been described as "a tingling, static-like sensation across the scalp, back of the neck and at times further areas in response to specific triggering audio and visual stimuli." People who enjoy ASMR visit YouTube channels that let them listen in on the sounds of people whispering, talking in a monotone, getting a haircut, or using a hair dryer. The Russian American performer Maria Viktorovna has been called the "queen of ASMR" because of the popularity of her Gentle Whispering ASMR YouTube channel. Her fans number in the millions. They find relief from depression, anxiety, pain, or stress in the hypnotic drone of her voice as she alternately whispers in Russian and English. Sometimes ASMR involves role-playing, where the viewer receives a virtual service, such as a shampoo or some type of medical exam, but the psychic rewards are chiefly derived from the timbre of the sound.

Though the mechanism of ASMR is not yet fully understood, it's clear that ASMR is rooted in biology rather than psychology. Exposure to certain preferred sounds reliably triggers changes in mood and physiology, such as reduced heart rate, deeper breathing, and a greater feeling of calm. ASMR fans emphasize that it is not sexual gratification they are looking for but, rather, the deep relaxation and "state of flow" that ASMR sounds elicit.

Though many sounds can provoke deep feelings of pleasure or comfort, sometimes they induce the opposite reaction. For a small percentage of listeners, commonplace sounds—sounds that most people don't even notice—can be as irritating as a shrieking baby. The term for this negative reaction to sound is "misophonia." People with misophonia experience revolting displeasure when they hear certain ordinary sounds like chewing, swallowing, or breathing. These so-called trigger sounds can spark anger, anxiety, and symptoms of panic. Oddly, people with misophonia are not especially sensitive to sounds that most of us consider annoying, such as the high-pitched whine of a dentist's drill or fingernails scraping down a chalkboard. Those noises evoke the same level of unpleasantness in misophonic individuals as they do in the rest of us. The key difference is that misophonia-inducing sounds are *interoceptive*: they are noises we make with our bodies.

Brain studies of people with misophonia reveal that when they hear one of their trigger sounds, they experience a high level of activation in the insular cortex, a deep-brain structure associated with emotion processing, particularly the feeling of disgust. If you want to fire up your insula right now, imagine the sound and feel of eating a juicy cockroach. Misophonia is thus the exact opposite mental reaction to timbre than the one Mr. B experiences while listening to Johnny Cash.

Personally, I find ASMR to be rather misophonic. Simply *thinking* about ASMR causes me to experience enough anxiety that I could not bring myself to click Play when researching ASMR videos on YouTube. In contrast, the animalistic bawl of bagpipes—despised by some as sour, droning, and monotonous—always gives me a warm surge of pleasure.

5

Of all the timbres we hear, none has the same concentrated effect on us as the sound of the human voice. The natural coupling of sound and emotions is rooted in choices faced by our ancient forebears. Responding to sounds didn't just keep the earliest humans out of danger. It helped them find love—or a lover, at any rate. Deliberate training can teach us to distinguish a Stradivarius from a Guarneri. But we don't need any training to know what kind of voice we find sexy.

Humans historically engaged in sexual activity at night, writes psychoacoustician Josh McDermott. The dimness of the forest or cave made it difficult to see your mate's face and body, so we evolved to have a high erotic regard for the sound of voices in the dark. For their part, female voices change their pitch as they proceed through the menstrual cycle and are therefore considered an "honest" indicator of fertility. Males tend to favor a female voice that is gentle and breathy—think Scarlett Johansson—and they instinctively associate this timbre with femininity. The attractiveness of a woman's voice (as perceived by men) is a strong predictor of a woman's sexual promiscuity (measured as the age of her first intercourse, the number of her sexual partners, and the number of times she cheats on a committed partner). As it turns out, women with sexy voices have more sex than women with sexy bodies.

Women show a bias in their evaluation of male voices, too— but this bias may not be based on actuality. A study examining the relationship between male vocal timbres and women's attraction to those timbres revealed that even though women exhibit clear preferences for certain voices, male vocal timbre doesn't always signify what we think it does. Voice recordings from thirty-four different men between eighteen and thirty years old were presented to fifty-four women in the same age range. The women were asked to listen to each voice and guess the man's attractiveness, age, weight, height, muscularity, and whether he had a hairy chest. Women exhibited very strong agreement across their guesses: the men with the deepest voices were rated as more attractive, older, heavier, more muscular, and as possessing more chest hair. In reality, there is no relationship between the timbre of a man's voice and any of his physical characteristics, except for weight. Researcher Sarah Collins pointed out that even though women may choose male sexual partners based on their vocal timbres, "the function of the preference is unclear given that the estimates were generally incorrect."

Record production is a profoundly male-dominated field, and so it was a rare occurrence at Berklee when I looked up during my production class one day and discovered that the only students who had shown up were female. I decided I needed to take advantage of such an unlikely moment. It was time for an all-girl chat. I launched into a question that I had often thought about but never actually asked other women: *Which male vocalist do you think has the most attractive voice?*

The young women were eager to share. One by one they pulled out their phones and dialed up the singers they found to be swoon-worthy based on vocal timbre alone. Most of the voices they responded to seemed to reflect their own young and confident outlooks. Ryan Adams: youthful, approachable, relatable. Jason Mraz: breathy, intimate, with a "boy next door" timbre. Miguel: also breathy and intimate, with an authentic, sincere, and naïve vibe. Elliott Smith: the breathiest of all, fragile, confessional. And one standout: Jason Aldean—deep and country strong, with great technique and impressive control. All the men these young women adored were talented singers, but their voices were far from my own timbral sweet spot.

Ever since I first heard him decades ago, I have loved the voice of Kevin Sandbloom. Listen to "Say Yes" from his 2005 EP *Delta*. You will hear a vocal timbre that I find so personally swoon-worthy, it makes me forget my own name. To my ear he sounds like a close-mic'd and very expensive cello, with the bow dragging slowly yet intently across the strings. An uncommon privilege to hear.

Each of the women in my production class—including me—enthusiastically professed a love for sexy male voices. This underscores the primacy of the vocal track in music. Other instruments are emotive, but only the voice gives us both emotional content and a sense of the performer's identity and physical fitness. Vocal range is a way for singers to hint at their sexual prowess, as we saw with Frank Sinatra in the Melody chapter.

Aside from falsetto, most men sing in their "chest voice," also called the modal or speaking register. In contrast, female singers tend to use their "head voice," especially early in their career. (These anatomical terms refer to where in the body most of the sonic resonance is coming from.) The pitch of our voices depends upon the length of our vocal cords. Interestingly, humans are one of the rare species to exhibit sexual dimorphism in our voices. Listen to a horse whinny, a dog bark, or a cat meow. You can't tell if the vocalizing animal is male or female. *Young* humans exhibit identical vocal timbres: boys and girls sound alike. But when boys go through puberty, the release of testosterone lengthens their vocal cords and triggers the development of the "laryngeal prominence," more commonly known as the Adam's apple. After puberty, the average male voice is pitched an octave lower than the average female's. This dimorphism makes the human voice *gender expressive*—we can use an adult's voice to infer something about their sexual appeal.

Voice training helps singers expand their range and shift seamlessly from their chest voice to their head voice and back again. This takes strength and control, so when we hear a man singing in falsetto, it sends the message that he has an extra gear and the power to command it. Likewise, when we hear a woman sing in a deep chest voice, she is broadcasting that she possesses a strength that most women do not. Nina Simone is the quintessential example of a female with a strong chest voice, as heard on "No Good Man," but you can also hear a deliciously deep female register from Miley Cyrus, Etta James, Tanya Tucker, the Russian rock singer Juliana Strangelove, and many others. Deep female voices are especially impressive because males typically have a wider pitch range than do females. It is easier for men to constrict their vocal folds and sing higher than it is for women to lengthen theirs and sing lower, so men can sound feminine more easily than women can sound masculine.

The complexity of timbre makes it the most individualized of all musical dimensions. Listeners may not have strongly biased opinions about melodies, lyrics, or rhythms but, as we've seen with ASMR, misophonia, Mr. B, and the preferences of the women in my production class, a person's reaction to any given timbre ranges from ardor to repulsion. We each have a highly personal constellation of sweet spots on the dimension of timbre, and this constellation is one of the most decisive ways your listener profile is unique to you.

Memory for Music

Music and memory are devoted mental partners. Our memory for songs and records is surprisingly resistant to decay, even in the face of physiological damage. Happily, even music lovers afflicted with mild to moderate Alzheimer's disease maintain their passion for music. Individuals with severe memory impairments can usually recall the melody, lyrics, and rhythm of songs, especially songs that were popular when they were young.

Linking words to melody "doubles up" brain activity by engaging both the left and the right auditory cortices. This "dual encoding" of memories across both sides of the brain may have contributed to the preservation of oral histories. Before *Homo sapiens* invented writing, ancient peoples *sang* their tales, myths, and chronicles to the younger generation. Melodies and rhyme schemes provided cognitive cues that aided listeners in remembering the words of a story—if you did happen to forget the words, the melody could help you remember them, and vice versa.

CHAPTER 8

FORM AND FUNCTION

This Is What It Sounds Like to a Record Producer

✧

Rick Hall lived with that record like a hermit
until he had it just right. He knew exactly
what he wanted, and he wasn't going
to stop until he got it.

—*Arthur Alexander, country-soul singer*

1

THE DOCUMENTARY FILM *MUSCLE SHOALS* EXPLORES THE
history of FAME: a legendary recording studio nestled in the north-
western corner of Alabama, near the Tennessee River. FAME was
founded by one of my record-producing idols, the late Rick Hall.
Rick packed the *Billboard* charts in the 1960s and '70s with records
by Aretha Franklin, Etta James, Percy Sledge, the Rolling Stones,
and more. In the film, the singer Candi Staton talks about Hall's
legendary perfectionism during sessions with the celebrated house
band, the Swampers, who contributed to many of his hits. Staton
describes how they'd work for days—*days!*—trying to get the performance right on a single song before he would be satisfied.

This persistence amazes me. Hour after hour, working on the
same track with top-flight musicians, Rick wasn't satisfied. Over and
over, take after take, his finger would press the talk-back button to
say to the Swampers, *"Again . . ."* With so much talent behind the
control room glass, why did it take so long to get a perfect take?
What was missing? More to the point, what was he *listening for*?

Rick wasn't listening for a great song. He wouldn't start recording
unless he knew he was working with a promising tune. He wasn't
listening for great performances, either. He knew that the Swampers
were capable of world-class execution. No, Rick Hall was listening
for a great *record*.

A record emerges when all its individual performances mesh into
a wondrous gestalt where, to quote Gestalt psychologist Kurt Koffka,
"the whole is other than the sum of the parts." Capturing music in

a storage medium is *recording*. Capturing music that resonates with a listener's soul is *record making*.

I'm changing things up in this chapter. Instead of helping you better understand what music sounds like to you, I'm going to help you hear what it sounds like to a record producer. My hope is that by taking you inside the minds of people who listen to music for a living, you might get more in touch with your own listener profile. I'll confess another private wish, too: I hope a few readers might be inspired by this chapter to consider a career as a record producer.

Listening to music is a different talent from performing music, just as directing a film requires a different skill set than acting in one. In the days before home recording software, successful producers often started out as something other than a musician or songwriter. Gus Dudgeon, Jerry Wexler, Nigel Godrich, Sylvia Massy, Keith and Hank Shocklee, Mark Ronson, and Boi-1da all felt the call to make records, while consciously forgoing the path to becoming a performing artist. Like others in the business, these justifiably famous record producers began their production careers as music journalists (Wexler), A&R executives (Dudgeon), recording engineers (Godrich, Massy), technical innovators (the Shocklee brothers, Just Blaze), or DJs (Ronson, Avicii). Two of my greatest heroes, Sam Phillips (Sun Studio) and Rick Hall (FAME Studios), were paragons of production, even though neither seriously considered a career as a musician.

Being a musician can certainly help with many aspects of record production. Often, though, it requires a readjustment of your listening skills. All my Berklee students are talented musicians. Each has spent countless hours honing their craft, day after day, month after month, year after year. Musical practice, especially listening attentively to individual tones, strengthens the auditory-processing circuitry and builds "auditory athletes." Just as a tennis player who practices every day develops motor skills and hand-eye coordination, a trained musician develops a capacity for rapidly perceiving and responding to subtle differences in all sounds, not just music. This capacity is called "analytic listening."

Musicians are trained to listen for intricate and subtle acoustic details in sounds. As a musician advances in her analytic listening ability, she learns to distinguish and produce tones that are on pitch and in time. She learns how to emphasize certain notes to help listeners feel the rhythm of a particular piece and how to alter the length of her phrases for emotional impact. She learns the nuances of coaxing the finest timbre from her instrument, when to use a soft touch, and when to dig in. After she joins a band, analytic listening helps her hear how her own musical gestures blend in with other musicians'. Berklee students are adept at adjusting their performances to complement, support, or build upon the performances of an ensemble. But part of what I do as a professor in the Music Production and Engineering Department is train students to listen like producers, and that means learning to hear music that, as Rick Hall put it, "has mass appeal to the common folks who make up the heart of the record-buying public."

Record producers need an ear for the holistic totality of *all* the sonic elements in a song. This is called "synthetic listening." A producer uses synthetic listening to distinguish among the details on a record that must be perfect from those details that might benefit from minor or even deliberate mistakes. Analytic listening ability develops through years of formal musical training. Synthetic listening, in contrast, develops through years of listening to records.

When my musician schoolmates were practicing their scales, I was listening to records. When my adolescent songwriter peers were putting their feelings into lyrics, I was listening to records. When my college-age musical cohort was rehearsing in bands and playing gigs, I was listening to records. When my contemporaries were touring concert halls, selling songs, and getting signed by labels, I was listening to records and studying the technical skills I would need to make them. When I finally had the opportunity to make records of my own, all those years of intense, active listening had provided me with a deep reservoir of knowledge and intuition about records that was carved into my neural pathways. Like all producers, I considered my mental musical library to be a prized professional attribute, one I relied on as I made rapid aesthetic choices in the studio, recording take after take of my own.

For great record producers like Rick Hall, music listening is as inseparable from their minds as their shadow is from their body. No matter where they find themselves, music is always the loudest voice in the room. For these producers, the term "background music" is an oxymoron. Their brains are *always* attuned to any music within earshot. Music can even make mundane activities difficult or impossible to focus on. I recently rewatched a dramatic pub scene on a television show three times before I could follow the dialogue, because an Al Green record was playing on the pub's jukebox. I simply could not wrench my attention away from the tune.

For veteran record producers, the basic experiences of reading, writing, conversing, cooking, exercising, and even (for me, at least) driving can be hard to focus on if music is playing. Our memories of waiting for a friend in a café, shopping for groceries, pumping gas, or lolling on a beach are shot through with the music that played nearby during these episodes. Success as a producer involves leveraging your listener profile, with all its sweet spots, to craft a tapestry of sound that makes other listeners say:

This is the music of me!

2

Often the first question a producer considers when working on a new record is: What is this record's *function*?

There is a crucial relationship between the form of a creation and the role it plays in the lives of its consumers. I think of this relationship as "the beanbag dilemma." A beanbag chair is a perfect example of a timeless creation of limited functionality. With its unusual form (no legs or back and a bendable seat), a beanbag won't be found in offices or dining rooms, though it's commonplace in children's bedrooms and family dens. The classic aluminum Navy Chair provides a sharp contrast. A Navy Chair has four legs, a stiff back, and a flat seat, and is used in a wide variety of settings from academic conferences to battleships.

The more unusual the form, the more limited the function. A record with a classic form like Taylor Swift's "All Too Well" can usefully function throughout a listener's day, from the morning commute to the after-work cocktail hour, and even the final hour of winding down before bed. Converge's magnificent if unconventional "Fault and Fracture" has a narrower utility. The more limited the function, the less the commercial appeal, though a record with an unorthodox form can sometimes become well regarded or even iconic, such as Queen's "Bohemian Rhapsody."

The lion's share of responsibility for a record's *form* is borne by the artists and songwriters who create the raw materials for the product: the songs themselves. Most of the responsibility for a record's *function* is borne by the producer. The producer must consider how the artist's songs are likely to be used, weighing ideas about the ideal audience, the ideal context, and the ideal listener response. Because musicians are artists, even among pros there is often a strong pull to make "art for art's sake"—and, often, we do—but the professional survival of both musician and producer ultimately depends on achieving some degree of commercial success.

When we work in avant-garde styles (such as noise pop or free jazz), we are well aware that, like a beanbag chair, our record has limited functionality. If we opt for a more classic form, then—as the novelty-popularity curve reminds us—the record will be easier to market because it will appeal to many more listener profiles and can be used in many more musical contexts. Nevertheless, highly functional forms present their own set of challenges, including the biggest challenge of all: the competition.

The top commercial producers make records with familiar forms. Consequently, the consumer always has a plentiful supply of such records to choose from. Competing in a small arena with fewer peers is easier than competing in a vast marketplace of highly similar objects, but most of the rewards in the music business are commensurate with the size of the market.

When a producer considers the functionality of a record, she weighs the ideal context in which it might shine. D'Angelo's "Untitled (How Does It Feel)" hews to the timeless form of make-out records: moderate tempo, uniform dynamics, crooning vocal, legato performances. Kool & the Gang's "Celebration" lets you know straight up that they want it blasted at any and all festivities: "There's a party goin' on right here . . ." Queen designed the famous stomp-stomp-CLAP rhythm for "We Will Rock You" not so consumers could listen to it alone in the dark but to engage the massive audiences at their arena concerts, where only a simple, slow beat lets eighty thousand fans perform in unison. As we learned in the Rhythm chapter, moving and singing as a group generates communal feeling and creates a memorable social event. (This is the function of "arena rock": allowing thousands of fans to bond with one another, as well as with the band.) Many songs by the Grateful Dead, such as "Playing in the Band" on the 1971 *Grateful Dead* album, were written to function as live concert performances that supported their fans' psychedelic drug trips. Consequently, some Grateful Dead songs are less appealing on records. As one critic put it, "[Even] the band's live albums don't come close to capturing the band's concert experience."

In my lifetime, I've witnessed more than one major shift in the functionality of records, but perhaps the most momentous was the broad-based cultural transition from active listening to passive listening. As with any product, changing how a record is consumed changes how it's made. In the early days of radio and turntables, most music consumers engaged in active listening by sitting at the radio or in the room where music was playing and giving it their full attention. In my own generation, kids would go to a friend's house to listen to records, usually toting along a few of our own. We'd spread them out in front of the stereo and troll through the front and back covers as though searching for clues to another world, sharing thoughts on what a lyric meant or what the artist might have felt or been doing when the song was written.

The halcyon world of active, communal listening was rocked by the arrival of the Sony Walkman in 1979. It signaled the beginning of a new era of portable, personalized record listening. For the first time, listeners could privately enjoy their favorite records in a broader variety of everyday settings, such as the office, the park, and even the library—contexts where we perform non-musical activities that demand our attention. This required *passive* listening: when you don't focus your attention exclusively on the music but simply want a background soundtrack to keep you motivated, relaxed, or connected while you do something else. Most music in the twenty-first century is consumed passively.

Record producers are aware that, more than ever before, their most vehement creative efforts may fall upon ears that are only "half-listening." Thus, contemporary producers must consider how much cognitive effort their record will require to be fully enjoyed. Highly novel or complex records are best appreciated through active listening, so that the small gems of harmonic layering or poetic lyrics don't escape notice. On the flip side, a record suitable for passive listening—and, therefore, targeting a broader audience—should employ familiar forms and fewer musical surprises.

3

Before any conversation about form and function can be had, the producer must get hired. The first step in getting hired is the producer audition.

The producer audition is a lot like a blind date between a producer and a band or artist. The goal is to learn if you are well matched—or if you might each be better off partnering with someone else. The producer sits down with the talent and exchanges views on the upcoming record to learn if what needs doing is in the producer's wheelhouse. During an audition, a producer must listen very attentively to the artist before she can hope to represent his creative vision. Experienced producers carefully consider what they hear before deciding what to say.

I once had a producer audition with Lou Reed, the legendary front man for the Velvet Underground. It was a rainy evening in the SoHo district of Manhattan, in a low-key and elegant Japanese restaurant. Lou was a bona fide rock god. I dearly wanted to work with him, but I could feel my spirits sag as he described the record he wanted to make. He wanted it to be an intense rock record with a heavy emphasis on improvisation and spur-of-the-moment inspiration. I surely love listening to that type of record, but I knew that I didn't have the ear to produce it. I am better suited to getting the little details just right than to capturing an extemporaneous project in one overarching gesture. I like to musically underscore a song's lyrics. I place great emphasis on the nuances of bass and drums. One of my favorite ways to experiment is through complex harmonic layering. These preferences and skills make me perfect for crafting stylized studio records, but not so perfect on a record where the main thing you're selling is the vibe.

As painful as it was, I had to confess that I wasn't the right producer for Lou's project. He ultimately decided to work with his former collaborator, Hal Willner. It was the right choice for both of us. I would never have been able to live with the shame of defanging a Lou Reed record!

If an artist decides they want to work with me, then one of the best ways for us to learn about each other is to have a record pull. What do *you* think is a solid groove? A great vocal sound? A knock-you-out lyric? Too much reverb? When you say you want to make a soul record, are you imagining Drake or Solomon Burke? When you say "classic country album," do you mean Patsy Cline or Kasey Musgraves? It is a terrible feeling to be in the studio on day one and realize that the snare sound you thought was totally perfect is the artist's idea of totally wrong.

Many years ago, I was enjoying a record pull with Paul Wester-berg of the Replacements a few weeks before I was going to engineer for him. I recommended the musician David Coleman (the younger brother of Lisa Coleman, the keyboardist in Prince's band the Revolution), known for co-writing the title track on Prince's album *Around the World in a Day*. David was a breathtaking young cellist who could play in a wide range of styles and was extraordinarily creative. Paul had only one question: "What kind of shoes does he wear?"

"Those flat, black slippers that you get in Chinatown for a dollar," I replied.

Hidden beneath this exchange was Paul's real question: "Is David an uptight, classically trained cellist who expects sessions to be planned and run like clockwork—or is he someone that a pioneer of alternative rock can stand to be in the studio with?" For Paul Wester-berg, Chinese slippers was the right answer.

It is a simple fact of both romantic courtship and record production that our very best efforts will not please everyone. If we want a successful record, we need to please only one broad category of audience. Fortunately, there are three to choose from: *critics*, *musicians*, and the *public*. Each group of listeners uses different criteria to judge a record. Thus, each audience bestows a different kind of reward on record makers. Before embarking on a new project, it's smart to pick one of these audiences to target. Though there are some records that appeal to all three audiences, this overlapping territory is small and *extremely* difficult to hit.

So difficult, in fact, that any record that manages to simultaneously earn the love of critics, musicians, and the public is considered to have worn the "Triple Crown."

Critics and scholars of music, like their colleagues in film and literature, are listening for ideas whose time has come. They ask, "Who is doing the kind of work that contemporary culture could use right now?" Critics assess music with more than personal taste in mind: they want to illuminate underappreciated talents, promote worthy young artists, and reward those who show daring or intellect. They are informed listeners whose job is to regard a musical artist in the context of similar artists and the arc of history, so that the listening public can make informed choices. Which of the leading artists in a new style will become legends? Why did that popular band's new record disappoint? Talented critics can even bend the arc in an aesthetically worthwhile direction. When a record pleases the critics, they publicize you, and so it may be said that the critics offer record makers the reward of *fame*.

Musicians seek inspiration and points of reference from records. They are listening for where the skill bar is set and for exciting ideas or techniques to emulate. We are never so harsh as when we're judging someone who reminds us of ourselves. Musicians listen and ask, "Could I do that?" If the answer is yes, they are less likely to be impressed. Based on what I've experienced at Berklee, many young musicians judge other artists in a very binary way: they either respect or dismiss an artist, according to their own intimate knowledge of what it takes to write, sing, program, and perform music. When a record pleases other musicians, they admire you and cite you as an exemplar to other musicians, and so it may be said that musicians offer the reward of *respect*.

The Triple Crown

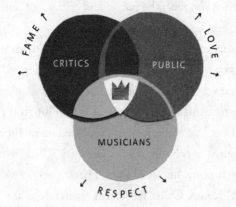

The *general public* is the best-known and most easily described audience. Unlike critics and musicians, the public isn't "betting on the races." They like what they like and don't seem to care who is the most clever, creative, or virtuoso. Instead, the public listens for music that lets them pay a relatively cheap price in mental effort to get a good-sized return in musical pleasure. Just as fewer people go to art films than to Hollywood blockbusters, most listeners tend to ignore a great deal of music that critics praise in favor of less demanding but more thrilling diversions. When your record pleases the public, they follow you, go to your shows, stream your music, wear your T-shirts, and call themselves fans, and so it may be said that the public offers the reward of *love*.

Years ago, when Tommy Jordan, Greg Kurstin, and I were having lunch during the making of Geggy Tah's *Sacred Cow* album, Greg posed this question: "Who has worn the Triple Crown the longest?" It was hard to think of a band or an artist who had worn the crown for longer than a single album. Michael Jackson came to mind, but many musicians assign the credit for his success to his producer, Quincy Jones. The public adored Led Zeppelin, but many critics despised that band. Musicians revered Jimi Hendrix as the world's greatest guitar player, but the public didn't embrace him the way they did Eric Clapton. The Beatles were probably the most popular band in their day with the public and critics, but musicians were more likely to emulate the Rolling Stones.

Greg Kurstin ended our lunchtime debate with two words: "Duke Ellington." Ellington was a bandleader during the height of the big band era who assembled what has been called the "best-known orchestral unit in the history of jazz." Ellington consistently attracted sold-out live audiences and recorded many hit records. He was (and remains) at the top of most jazz critics' lists of bandleaders, even winning a posthumous Pulitzer Prize Special Award in 1999 for his musical genius. Other musicians revered him as a god on the piano. In the history of American music, Sir Duke may have worn the crown the longest.

Prince also managed to wear the crown for a brief time. As he built his reputation, he had the foresight to recognize that different audiences listened through different filters. To advance his career, he made the daring tactical move to record albums that targeted different audiences. Prince's first two albums did what most beginners' early records do: they showed off his craft in the style of the day (although sprinkled with enough innovation to hint at what he was capable of). His first single, "Soft and Wet," from his debut album, sounds not that far removed from the keyboard-driven R&B dance records that were popular in the late 1970s. But for his third album, he made the risky choice to abandon the soul music fans who'd just begun to give him support. That shocking third album—*Dirty Mind*—was beyond the pale for R&B radio with its taboo-busting themes. Yet its combination of soul and punk ignited the interest of the one audience that could give him the necessary attention to compete at the highest level: the critics. The title track, "Dirty Mind," had more in common with new wave pop/rock music than it did with R&B music, which, in the early 1980s, was still finding its way out of disco. It worked. Critics from New York to Los Angeles trumpeted Prince's stylistic maneuver as evidence that "he is heir to the defiant rock-and-roll tradition of Elvis Presley, Mick Jagger and Jimi Hendrix."

For his next album, *Controversy*, Prince targeted another audience: musicians. He played nearly every instrument himself on this tight, cohesive, and, ironically, less controversial record. Though many older musicians still regarded him as a bit of a freak, they had to acknowledge that he was a phenomenal performer on keys *and* guitar *and* bass *and* vocals *and* . . . the young artist could write a hook. The song "Private Joy" shows off his practically unparalleled ability to write joyous, infectious pop. But if you home in on the bass playing, keyboard technique, guitar solo, background vocal harmonies, and stunning vocals, you hear a staggering range of raw talent.

Now that Prince had attained good press from the critics and respect from musicians, it was time to target the public. Once again, it worked. The album *1999* gave Prince his first crossover single on the pop charts with "Little Red Corvette."

His sixth album was *Purple Rain*—a global mega-hit that enabled Prince to don the Triple Crown. Listen to "Purple Rain" and imagine hearing it in 1984 through the ears of each of the three audiences. You might agree that this song, like the entire *Purple Rain* album, shows innovative artistic merit, impressive musical talent, and—most important for a great many listener profiles in the record-buying public—inhabits the Goldilocks zone of "just right."

4

After a producer has landed the gig, she and the artist do some preproduction before heading into the studio. In preproduction, songs are rehearsed over and over again as parts are added, discarded, and changed in hopes of finding the record's ideal form. A bonus of this demoing process is that the producer can get acquainted with the performer's playing and singing technique, temperament, and creative flexibility before stepping into the high-pressure environment of the recording studio.

The producer must be keenly attuned to the global whole, listening synthetically to the blend of elements and considering where—head, heart, or hips?—the spotlight of attention should be aimed to suit each particular song. Would this song work best as a danceable record? A songwriter's showcase? An emblem of sexuality or intellect? Can we find a harmony that provides counterpoint to both the melody and the lyrics? What timbres would show off its strengths and shore up its weaknesses? In preproduction we can try out different options for songs while simultaneously learning about our artist's strengths.

Here is an example from my own studio days with the Cuban American artist Nil Lara. The song "Baby" needed to showcase Nil's two greatest artistic assets: passion and power. The guitar timbres on this record are robust and sharp, chosen to convey strength. The sound of the bass and drums are present and "in your face," as we say, for power. But we want listeners to relate songs to their own lives, and so it is important that an "in your face" approach sometimes backs off a little. To achieve that, the guitars change to a slightly softer tone where Nil pleads "Be my savior." His voice drops back from the mic to make it feel less personal and more universal.

In preproduction, we often try out different tempos to hear how the melody works both fast and slow, learning something new about the song in the process. We saw in the Melody chapter that Pharrell Williams's "Happy" works best at a moderately fast tempo (appropriate for a song about happiness). But there doesn't always have to be congruence between tempo and lyrics, as we saw in the Lyrics chapter with Train's "50 Ways to Say Goodbye." The basic form of Train's song (its melody, lyrics, and rhythm) could be adapted to fit a slower tempo. If a slower version of the song abandoned the mariachis, this record would become a more traditional tale of romantic woe.

During preproduction, we experiment with rhythm to see if it sets the pace better when it's more staccato or legato. For example, the rhythm for a song about flirting is congruent with the lyrical message when delivered in a brisk, skipping groove that mirrors the singer's playful self-confidence. Alternatively, you could add a more deliberate, sexually persistent subtext by deploying a straighter rhythm. We heard in the Rhythm chapter how syncopation emphasizes the "hips" on a record but also adds some bounce. This "bounce" risks detracting from the unwavering forward momentum on make-out records like D'Angelo's "Untitled (How Does It Feel)." Its 6/8 rhythm (doubling up the traditional waltz time of 3/4) has a straighter feel than a flirtier record that might employ more syncopation. D'Angelo's kick drum lands predictably on the one and the rimshot lands predictably on the four in every bar of this seven-minute seduction. Syncopation can sound like the *prelude* to a romantic dance, while straighter rhythms can do a better job of implying the dance itself.

In 1999, I made an album with blues guitarist Robben Ford featuring the song "Don't Lose Your Faith in Me." It was a plea for forgiveness and a second chance. The drums, played by the incomparable Vinnie Colaiuta, are steady and even-toned, never distracting us with unnecessary dynamic flourishes. The rhythm section gives Robben the space to plead his case. Robben is a blues/jazz guitarist who was hailed as a wunderkind when he emerged on the scene, but on this album, his vocal prowess wasn't as practiced or as fluent as his guitar technique. That's okay. When a lead singer is less than a powerhouse, another instrument can deliver the passion.

Had we made the choice to record this song with electric guitar instead of acoustic guitar, a fervid guitar solo would have added the necessary ardor, but that would have been less than ideal for this song. Robben's voice would've had to work harder (at the level we heard on "Baby" with Nil Lara) to cut through a timbral forest of guitars. Therefore, the job of delivering passion went to the renowned composer and arranger Roger Kellaway, whose remorseful strings supply the backstory underlying the lyrics. Robben's voice loses its power in some places, but that doesn't matter. The musical backing of Kellaway's strings endows the vocals with sincerity.

Vocally, Robben means what he says—and what he can't say, his band says for him. A critic writing in the *Washington Post* observed, "Ford often seems more interested in making the lyrics count than in demonstrating his guitar skills. While he's hardly a commanding vocalist, he infuses most of the songs here with sufficient emotion to sustain interest until his guitar asserts itself." This was our intention on the album, and I was glad to know that our message was successfully received by at least one listener.

The artist, not the producer, has the final say regarding what stays and what goes. (There are exceptions for so-called industry bands— music-executive creations put together expressly for the purpose of attracting tweens, such as New Kids on the Block, the Spice Girls, and BTS.) If a producer disagrees with the artist's choices, each producer must decide for himself how far he's willing to go to make a point. Producer Tony Berg once explained to me why he capitulated and let the band we were working with choose a direction that went against his better judgment:

"I might make two hundred albums in my career. They might make two."

5

So how does a producer decide what a record should sound like? Perhaps the most important directive I give my students is this: *Grow* the seed, don't *bring* the seed.

There's a common misconception that a producer's managerial style should be authoritarian, calling the shots and firing the laggards in order to grind out the record that the producer deems best. Dictatorial direction may have been true in the early days of record making, but that's not how it works today. The producer's job is not to apply a predetermined sound to a particular song like a coat of paint—you don't copy an existing record just because it was a hit. Nor is the job to record the skeletal bones of a song with no embellishment whatsoever. Rather, record producers listen deeply for the spark of creative intentionality and use the light of this spark to guide musical partners down the road ahead, offering alternate routes if needed, and always keeping a lookout for signs that we should rethink our destination.

Today's producers are creative partners who collaborate with gifted songwriters and artists. GRAMMY-winning producer Greg Wells visited Berklee in 2015 and spoke with the Music Production and Engineering students. He uttered a sentence that is profoundly true, yet seldom acknowledged: "You guys, sitting here today—you have no idea just how good *good* is out there." Performers who reach the top of the charts and earn the highest level of praise are, nearly without exception, even more talented than you might guess from their records. Once you've heard extraordinary musical talent, you never forget it.

Given the sublime level of talent available to many producers, you might think that the producer's objective is to bring out the most virtuoso performance possible from your artists. That's a rookie attitude. If you want to make the very best *record*, you shouldn't always opt for the very best a performer can do—you should opt for the *right* thing to do.

Listening for the "right" performance and sounds is perhaps the most advanced skill in a producer's repertoire. All our notions are educated guesses built from listening to other records, a strong sense of how listeners respond to music, and our own listener profile. A decent producer learns to hear when a part or performance or timbre just won't work. Once again, questions are asked. Is the problem the part or how it's being played? Is a thin timbre playing a robust role? Maybe those roomy, powerful drum sounds are working against you rather than for you. Perhaps the song itself is the problem. You mustn't talk yourself into liking something that deep down you suspect is flawed—otherwise, as Prince used to say, "It'll haunt you."

Even subtle performance gestures—deciding where the strongest accents should occur, lengthening a melodic phrase, inverting a chord, delaying the vocal by mere milliseconds, playing the snare with brushes instead of sticks—can have an outsized impact on the overall perception and function of a record. Producers find a way to hear the totality of these combined gestures and interpret what the performance as a whole is communicating. Sam Phillips listened for "imperfect perfection." His ability to hear it was perhaps his greatest talent. In Peter Guralnick's biography of the legendary record maker, he describes how Phillips would record performers in the studio:

> Some he would record with a delicacy of touch, others with the same kind of over-amped drive that he had discovered with "Rocket 88." But all [of his records] would reflect to the fullest extent that he was capable of the unmodified fulfillment of their inspiration, all would reflect the circumstances of the moment that gave them birth. And all that was required of him was the quality of "transference" that enabled him to put himself in the place of each person who stood in front of the microphone.

That "transference," or being able to imagine what it sounds like to the performers as well as to the audience, is the art of listening like a record producer. A timbre change, a shift in dynamics, inverting the chord, dropping some words to smooth out the melody, rearranging the song structure: all these possibilities twirl like colors around a barber pole until the performance gets close enough to the sweet spots on your listener profile that you can say those magic words.

I think we got it.

6

Tommy Jordan is a prolific songwriter who says that writing songs is like having sex: easy, fun, and something he would happily do every day. He adds, however, that making a record is more like raising an infant. It keeps you awake at night, drains your bank account, and all your love and care won't stop it from spitting up on you.

But look! You (the producer) and your partner (the artist) have done it! You've raised a bouncing, beaming record, and now your little darling is ready to go off to preschool and mingle with all the other tykes.

You and your partner couldn't be prouder of how your creation turned out. But now you feel a growing sense of unease as you wonder how your bundle of joy will get along with everybody else. Some of her quirks might need explaining, and perhaps that adorable way she has of barking like a dog instead of saying no will be annoying, not charming, to others. You and your partner drop her off at day care and sneak up to the window a few hours later to check in on how she's doing. Chances are, your kid is throwing a tantrum, putting paste in her hair, staring blankly into space, or otherwise behaving in a way that lets you know she's not quite the superstar you'd imagined. But every now and then you get lucky. You raise a kid who grows up, makes her mark in the world, and eventually takes care of you by returning some of the energy you put into her in the form of royalties.

One of my kids did exactly that.

Barenaked Ladies' *Stunt* album was produced by me and my fellow producer David Leonard, along with BNL, who received co-production credit. The band approached me in December 1997 about the possibility of working together, but I had to turn them down. I had only three weeks free in the new year, between previously booked projects. Not nearly enough time to do a full-length album, which in those days usually took eight to twelve weeks. To my surprise, they said that they had songs ready to go and could make a lot of progress in three weeks. Would I still consider working with them? If we ran out of time, we could hand off the ball to a second producer/mixer (David Leonard). Deal! It would be an intense three weeks, but through mutual friends, I knew Steven, Ed, Jim, Tyler, and Kevin to be smart and hardworking. I anticipated that it would be a joy.

And it *was* a joy, but they saw to that. I was unaware that the band had just been through a period of internal struggle and self-doubt. Their interpersonal dynamic—a delicate house of cards for every band—was changing and, as with any partnership, the time had come for some renegotiation. Their live concert album *Rock Spectacle* had just sold well in the United States, so if we could deliver some radio-friendly singles, their popularity could ascend to the next level.

Up to this point Steven Page had been the main singer-songwriter, but Ed Robertson was coming up fast as a writer and vocalist himself. There would be fewer "Steven songs" and more "Ed songs" on this album, and that changed the band's power balance and BNL's sound. To their credit, they had already worked out a new dynamic before I flew up to Scarborough, Ontario, to do four days of pre-production at their rehearsal facility. Stories of bands fighting and breaking up during the making of a record are legion, and it is a testament to BNL's maturity and good sense that they had hashed out their issues before the creative work began. The band members' camaraderie and commitment to one another was vital if we were to make real progress in just three weeks. Fortunately, I witnessed nothing but professionalism from the band—including when they stripped off all their clothes in the studio and recorded a song naked, as they did on every album.

First things first: What did BNL want to achieve with this record? They shared two goals with me. Barenaked Ladies was well known and well lauded in their native Canada but hadn't achieved the same level of success in the United States. In addition, their audiences tilted female. Their first goal was to make a studio album that would do well in the States. Second, they wanted to attract more male fans.

To achieve the first goal, we needed to address the rhythmic form of their music. To my ear, music on the Canadian charts in the 1990s reflected a greater influence from the British Isles than did music on the US charts. In the style of rock music popularized by the Beatles, the Kinks, and Queen, drummers often "tag" the melody line by playing a tom fill to accent words that end sections. This gives rock music a dynamic profile. Verses have moderate energy and choruses get much bigger, with clear segues from one section to the next. American charts reflected the influence of soul music and R&B, where this dynamic is less common. Soul, funk, R&B, and styles that evolved out of the blues tend to build tension not by rising and falling across sections but by keeping a groove and building steadily toward a climax at the end of the song.

To sell Barenaked Ladies "south of the border," we needed the drummer, Tyler Stewart, to obey the instruction Prince often gave his band: *Don't move.* This meant resist the urge to build dynamics with tom fills and just stay in the pocket as we change sections. This change in form means the rhythm section puts a solid but less distracting frame around the vocals. Tyler is a powerhouse drummer with one of the deepest rock pockets I've ever heard. Could he be less Kenny Aronoff (John Mellencamp, Jon Bon Jovi) and more James Gadson (Bill Withers, Marvin Gaye)? He could.

To achieve the second goal, we needed more tough, gritty guitar tones while keeping a feminine ear on the vocals. Steven and Ed are two of the cleverest songwriters (and people) I know. Their onstage repartee was second to none. But we needed to pick up a male audience, which meant that we'd have to be okay with serving up more attitude. Conventional wisdom says that guys have a different sense of humor than women do—the Three Stooges and "Dad jokes" are cases in point—and men will approve of a clever put-down in a way that might strike a woman as being mean-spirited. I kept my ear on the overall lyrical message of the album. Some of the darker lyrical themes would have to be framed in a way that was more direct, less cloaked by wit. Their female fans—and I was one of them—loved Steven and Ed for the impression of them we gleaned from their lyrics: funny, smart, sensitive, respectful, and, above all, approachable.

I arrived in Canada with demos of fourteen songs, but right away Ed asked, "Hey, did we send you the latest one? It's called 'One Week.'" Steven later said of the song that it was radically different from a typical BNL song but very much like their live shows. The verses borrow from hip-hop and reflect the segment in every show where Ed and Steven would freestyle—improvise lyrics against a backbeat. This needed some consideration. It was one thing for five white guys from Toronto to freestyle in concert, but another thing altogether to rap on a record. We had to stay true to ourselves: make it clear that this was a pop/rock band borrowing a signature element from hip-hop rather than appropriating an entire style.

I hadn't received the song's demo before my flight, so they ran through it in rehearsal. Ed played "One Week" on acoustic guitar with Jim Creeggan, Kevin Hearn, and Tyler accompanying on bass, keyboards, and drums. It was clear that the rhythm section should remain acoustic (rather than programming a drum machine and using keyboard bass), but the rhythm guitar needed to feature a clean electric tone, as in R&B. This would be a realistic-sounding record evoking the live show and capitalizing on the recent success of their live album. Although in principle it could have worked as a singer-songwriter vehicle, to pick up that male audience we needed to keep the tempo energetic and driving. Tyler, like any seasoned pro, gave each section—verse, pre-chorus, chorus—its own straight beat. Rather than accent with toms, he let his cymbals do the talking. He resisted the urge to forecast the upcoming section with a tom fill. Just as in R&B, this "straightness" gave the record power for what it *didn't* do—for what it held back.

Head, heart, or hips? Clearly this song's best feature was its damn clever lyrics. We realized that if we could counterbalance the fast, percussive rap vocal with a straight, heavy kick drum and a fat bass tone, it would lift the rap up into the spotlight of attention while anchoring it with power. I had learned with the success of Geggy Tah's "Whoever You Are," a year earlier, that rapid-fire lyrics delivered in a singsong melody appeals to kids. The "One Week" lyrics are kid-attractive, referencing Aquaman, Sailor Moon, and Samurais, but the topics are adult, addressing tantric sex, golf clubs, and Kurosawa films. Ed took the verses while Steven's distinctive tenor sang the more melodic pre-chorus and chorus. The arrangement showed off the entire band, making it an ensemble piece rather than a solo act.

I've always loved the timbre of a giant tone coming from a small place, and we wanted to add a part that sounded like someone playing along with the record. We fed Ed's guitar into a tiny Pignose amp, just a few inches tall, and close-mic'd it for the effect. The little bits of noodling throughout the track supplement the subtext of improvisation, in the same spirit as the freestyle verses.

I remember recording and "comping" Ed's vocals (comping is the immensely time-consuming process in which you pick the best lines from each vocal take to create a single lead vocal track) and thinking that we'd gotten the performance we needed. Tyler listened, then turned to me and said, "He can do better." Tyler pointed Ed back to the vocal booth. Tyler was right. Ed's additional performance was amazingly tight. Steven's vocals were quick and easy to record, as they usually are with veteran studio singers of such high caliber. Steven and the band couldn't resist a couple of inside jokes as the outro faded—"It'll still be two days till we say 'wasabi' / Birchmount Stadium Home of the Robbie." Usually I don't like inside jokes on a record because I think it can feel a bit cliquish, but the spirit of "One Week" seemed to justify it.

Six months later, I was sleeping in a hotel room in Sydney, Australia, when the phone rang. It was my manager, Sandy Roberton, calling from Los Angeles. He uttered the sweetest words a producer can hear:

"Congratulations! You've got the number one song in the country this week!"

Out of the hundreds of "little darlings" that I helped deliver and raise, this one had made its mark upon the world. I used the royalties from "One Week" to pay for my undergraduate education at the University of Minnesota, where I enrolled at the youthful age of forty-four. Fast-forward twelve years. I was now a professor lecturing on technique in front of a production class at Berklee. I mentioned "One Week." One of my students, Andrew Sarlo, leaned on the kid next to him and sang, in his deepest baritone, *"It's been . . ."* mimicking Steven Page's opening line. The two of them half-fell out of their seats, choking with laughter. When he could breathe again, Andrew explained that he'd been a kid in the schoolyard when "One Week" was on the radio. He'd loved the record. He and his friend would launch each other into hysterics by sidling up and singing *"It's been . . ."* into each other's ears. My prediction that schoolkids would love "One Week" was confirmed right there in the Berklee classroom.

This is the endless cycle that is the life of music: a kid in the schoolyard is captivated by a record. He listens more deeply and attentively than his classmates. Enchanted, he finds and listens to more and more records, each bringing him closer to his personal listener profile. In time, he parlays his musical skills into an admission to Berklee, where he continues to listen to as many records as he can while acquiring the skills of production. He graduates and is sought after for his talent, his personality, and, especially, his *ear*. After leaving Berklee, Andrew Sarlo went on to produce records of his own, eventually working on GRAMMY-nominated albums by Big Thief, Bon Iver, and Courtney Marie Andrews. Surely someone listening to one of Andrew Sarlo's records will go on to produce the records of tomorrow.

Listeners are an essential part of the endless cycle of music because all music makers start out as listeners. Out of that listening are birthed singers, dancers, performers, composers, DJs, record executives, technical innovators, sound designers, and record makers, all eager to show the next generation, *This is what it sounds like . . . to me.*

The Future of Music

Shimon is a four-armed marimba player created by the Robotic Musicianship group at the Georgia Institute of Technology. Sophisticated machine-learning algorithms enable Shimon to acquire music theory and use his knowledge of music to jam along with human performers in styles ranging from chamber music to dubstep.

In 2015, Shimon joined several accomplished jazz musicians onstage at the Kennedy Center in Washington, DC, for a truly impressive performance. Shimon managed to hold his own amid the talented ensemble. He demonstrated human-like awareness of his fellow players by orienting toward the soloists. He even expressed his enthusiasm for the performance by bopping his head along with the groove.

Will music created by artificial intelligence downgrade the art form? My own attitude sides with the listeners who love it: whenever music delights our sweet spots, then who can say it is inferior to any other music we find rewarding?

You can watch Shimon's Kennedy Center performance through a link on our website, ThisIsWhatItSoundsLike .com.

CHAPTER 9

FALLING IN LOVE

The Music of You

✦

It was all I thought about, all I lived for,
all I completely loved.

—*Miles Davis,*
speaking about his relationship with music

1

I STILL REMEMBER THE FIRST TIME I SAW HIM. IT FELT
like seeing a part of myself that I hadn't known existed. It wasn't so
much how he looked. In fact, he was a ragged mess. Tall and slender,
with long hair that hadn't seen a comb in a while, cigarette butt
dangling from his left hand. Clothes too big, shoes too worn. He was
standing at a chalkboard in a small, weakly lit room, holding a stump
of chalk between two fingers, one of which was missing its final joint.
Three men, all older than he, their bulk extending beyond the edges
of the flimsy folding chairs holding them up, leaned forward with
their elbows on their knees, staring mutely and uncomprehendingly
at the chalkboard. He was explaining how a television works. He
sketched the circuitry that carries the signal leaving the broadcasting tower to ultimately emerge as a picture on a screen.

My roommate had brought me to this hole-in-the-wall recording
studio to show me where she had recently taken a job in hopes of
meeting and marrying a rock star. She didn't seem to notice the
young engineer at all. Why wasn't *she* transfixed by him? Couldn't
she see that he was *perfect?!* I could barely breathe for feeling that I
had just encountered someone who would matter tremendously to
me in the years to come.

I can vividly recollect my love-at-first-sight moment. Chances
are, you've had some of your own. You see someone across the room
and—everything *stops*. You simultaneously freeze and go limp. You
lock onto them and trace their features with your eyes. You try to

summon the nerve to introduce yourself. What is going on? Why does *this* person provoke such resonance, but not any of *those* people? Is it their clothes, their scent, their eyes, their body language? Is it something they said? It's hard to describe, but you know it when you feel it.

Science has filled entire bookcases with explanations of quarks and black holes, yet the phenomenon of love at first sight remains largely unexplained. We do know that love at first sight is not based on intimate knowledge of another person. Just the opposite, in fact. The biggest factor seems to be exactly what we think it is: *instant physical attraction*. With a single glance, you feel an uncanny magnetic shove pushing you closer. Why does this person light up your private psychic caverns in a way that feels both intimately familiar and wholly unexplored?

Love at first listen has a lot in common with love at first sight. Falling in love with a record follows a similar process of instant attraction, coupled with the peculiar cognitive dissonance of feeling as if you've known the music forever. Every record maker yearns for this reaction—the unquenchable feeling that *I must have that!*

In terms of their neurobiology, we saw in the Novelty chapter that there is a difference between "liking" and "wanting." Liking is a simple hedonistic response. We can postpone going after something we like. But wanting is more powerful because the object of our desire feels *essential* to our well-being. Love at first listen arises from beneath our conscious considerations, from a deep wellspring of private attraction that even the most passionate music lovers may struggle to explain. But you can examine that thrilling, seemingly inexplicable feeling through something concrete, comprehensible, and scientific.

Your listener profile.

2

The qualities of every record you hear can be mapped onto the seven dimensions of your listener profile. The four musical dimensions—melody, lyrics, rhythm, and timbre—are each analyzed in a distinct brain network specialized for music. The three aesthetic dimensions—authenticity, realism, and novelty—are each processed by several interconnected higher-order brain regions that receive inputs from the four music-specific networks.

Every one of these seven dimensions can independently zap you with its own jolt of pleasure when you're listening to music. Some dimensions are capable of providing you with greater pleasure than others, though these "preferred dimensions" vary from person to person. One person might listen for melody, while another will dismiss any record that doesn't have a strong groove. One listener enjoys only familiar music genres, while another doesn't care if it's old or new as long as the lyrics are poetic. One fan loves computer-aided sound design, while another craves old-fashioned engineering. When a record's music matches up with one or more of your sweet spots, it stands a good chance of igniting love at first listen.

The sweet spots on your listener profile were formed out of genetic predisposition, cultural influence, and all the random and purposeful listening episodes you experienced over a lifetime of exposure to music. Most of your sweet spots started out broad and fuzzy in infancy, but as your musical world progressed from your first lullaby to your first live concert, they gradually became increasingly fine-grained. Your brain became more attuned to recognizing whether a melody, lyric, rhythm, or timbre was especially rewarding. Similarly, you unconsciously learned to listen for the kinds of authenticity, realism, and novelty that had delighted you in the past.

Every brain is formed out of an uncountable number of genetic and physiological events influenced by chance. Every brain grows up in a different musical culture. Every brain's trajectory through life is one of a kind. Together, these grand developmental truths derived from the science of individuality mean that your own listener profile, including its combination of preferred dimensions and sweet spots, is unique to you.

That's why I always tell my students: *Never be a music snob.*

Your taste in music is every bit as valid as mine. It is the limitless diversity of listener profiles that fuels the infinitely rich art form we love. If we all possessed identical sweet spots, then commercial records would thrill us all equally, but the art of music would be stagnant and homogeneous. Fortunately, our sweet spots are diverse. The particular pattern of rewards that you happen to experience from music doesn't signify your level of cultural sophistication or intellectual attainment. Some people wear only blue jeans, while others have never owned a single pair; whether someone likes denim says *nothing* about that person's fashion IQ. It just says that jeans suit them. The variety of human tastes makes life, and music, wonderful.

The graphic below maps out a fictional listener's sweet spots. Let's call her "Val." When examining Val's listener profile, recall that the four musical dimensions are not binary but consist of several different axes, each of which can have a sweet spot. For ease of explanation, in the graphic portraying Val's listener profile, I'm going to show the "narrow interval versus wide interval" axis for the melody dimension, the "personal and intimate versus general and philosophical" axis for the lyrics dimension, the "straight versus syncopated" axis for the rhythm dimension, and the "acoustic versus electronic" axis for the timbre dimension. Keep in mind that each of these musical dimensions is composed of multiple axes for multiple sources of listening pleasure.

Let's examine how a specific record interacts with Val's listener profile: Lil Nas X's "Old Town Road (feat. Billy Ray Cyrus)."

Because Lil Nas X was young and inexperienced when he wrote and sang this track, there is some below-the-neck naïveté in his

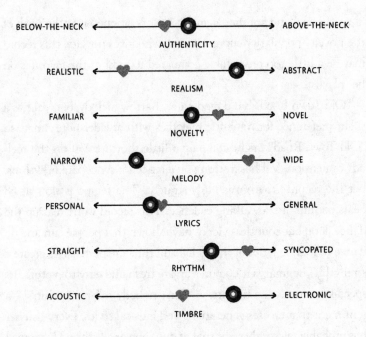

The Listener Profile

voice. The inclusion of Billy Ray Cyrus, on the other hand, adds years of polished technique. The production style strongly suggests that this was no "shoot from the hip" record but was the product of a good dose of above-the-neck skill. Lil Nas X's low-fidelity, home-studio techniques were carefully groomed by studio pros. On the aesthetic dimension of authenticity, the record lands in the middle of the axis. It may be a little too "clean" for Val, who prefers a sound that's more raw.

The modern digital production on "Old Town Road" renders it a more abstract than realistic record. The drums come from a machine, but the banjo evokes a real player, despite having been sampled and inserted into sections. There are plenty of vocals to give Val, who loves realism, ways to visualize the performances, yet this record is far from her sweet spot on the realism dimension. As for novelty, "Old Town Road" was a global smash in 2019, with a balance of novel and familiar elements that positioned it at the peak of the novelty-popularity curve. For Val, who loves novelty, this record may be a little too predictable compared to most of the favorites on her playlist.

"Old Town Road" has a moderately narrow melody, but Val has a slight preference for romantic melodies with a wider note range, so "Old Town Road" might come up a little short for Val on the melody dimension. Val has a strong preference for syncopated rhythms, but the record's rhythm is fairly straight. The tempo is slow at 68 beats per minute, a walking cadence. This record won't tear up the dance floor, but countless videos have shown that people can and do dance to it. The lyrics of this delightful tune are perhaps its greatest strength. The images it conjures up are fresh and easy to picture. Its special feature is the tag line—"Can't nobody tell me nothin'"—a sentiment that crosses generations and musical styles. Every listener has probably said or thought that at one time or another. The appeal of this record's lyrics is undeniable. When it comes to lyrical depth, this record is close to Val's lyrical sweet spot. The timbres are modern, electronic, and feature two charismatic male singers. The blend of banjo and 808 kick drum could be just right for Val, who likes hearing both acoustic and electronic sources on a record.

We can safely predict that "Old Town Road" wouldn't be Val's "love at first listen" record, but she would probably like it well enough to enjoy hearing it whenever it is within earshot. We can also see that because "Old Town Road" maps onto the center of several dimensions, most listeners will likely feel some connection to the song—a prediction born out by its record-setting global streaming numbers.

The ability to map out your own listener profile and decipher "the music of you" lies entirely within yourself. Only you can ferret out the traits and nuances of the records you love and figure out why you value them above others. Exploring your musical tastes can be a journey of self-discovery every bit as revelatory as pursuing a relationship. The best way to find out who you really are, deep down past artifice and the pressure of social approval and the mask of self-imagery, is to dive into your playlists . . .

. . . and listen.

3

Daydreaming is a profoundly undervalued gift. People who are perceived as spending too much time letting their mind wander are often denigrated as lazy or distracted, especially as adults. Cultural profiles tend to portray adult daydreamers as childlike for shirking their responsibilities and not being present for others. But what if there were far-reaching benefits from mental meandering? That seems to be the conclusion of a recent flood of research into the neuroscience of mind-wandering.

Whenever your mind works on a goal-directed task, such as composing an email or planning dinner, specialized task-specific networks in your brain are engaged, such as your "visual recognition network," "navigation network," and "decision-making network." These goal-pursuing networks govern *how we get things done*. But recently, scientists made a new discovery. When researchers ask people, "Are you thinking about something other than what you are currently doing?" respondents answer "yes" 30 to 50 percent of the time. This prompted neuroscientists to wonder what, exactly, was happening inside the brains of people whose minds were wandering rather than focusing on a task at hand.

A little more than a decade ago, scientists began to answer this question when they uncovered a previously unidentified brain network that turned on whenever a person was *not* actively doing something. Whenever we are preoccupied with daydreaming, fantasizing, or reflecting on ourselves, this strange network gets busy. Neuroscientists initially dubbed it the default mode network, using a bland descriptive term to suggest that the network was responsible for generating a sort of baseline state for the brain. Because the default network is most active during spontaneous, undirected thought, we might refer to it as the *mind-wandering network*.

Early investigators of the mind-wandering network believed it was half of a brain-spanning binary system: they assumed that the network turned on whenever we began daydreaming and shut off whenever we engaged in a goal-directed, externally oriented task. More recent discoveries have shown that this assumed dichotomy is only partly true. The mind-wandering network is activated not only when our minds are "idling" but when we are thinking creatively, such as trying to come up with an idea for a song.

Creative thought is a dynamic process. During creative thought, we shuttle back and forth between spontaneous thinking and more analytical evaluation of these spontaneous ideas. During the mind-wandering phase of creative thought, a random image, such as an oak tree with initials carved into its trunk, may pop into your head. During the analytical phase of creative thought, you consciously consider what you might *do* with your spontaneous thought, such as getting inspired to write a love song about hearts carved into trees.

Scientists have unearthed another revelation about the mind-wandering network, one even more relevant for explaining why we fall in love with a record. Our brain treats listening to music as a special form of daydreaming. When we're immersed in the enjoyment of a favorite record, our mind-wandering network lights up like fireworks. This observation helps explain many of the mysteries of our deep bond with music.

When your brain is idling, the contents of your reveries—what the psychologist William James labeled "flights of the mind"—contribute to your conscious conception of your personal self. Whenever you daydream or fantasize, your mind drifts to places that are intimate and private, thinking about what you like, what you need, and what you desire. Thus, when you listen to music you love—music that aligns with your sweet spots—you activate the part of your mind that fuels the deepest currents of your identity.

The connection between the mind-wandering network and aesthetic pleasure was first discovered while exploring viewers' reactions to paintings. Scientist Edward Vessel and his research team presented unfamiliar artwork in a broad range of visual styles to participants in an fMRI scanner. Because the participants had not seen the paintings before, they could not judge them based on previous experience or cultural reputation. Participants were asked to look at the paintings and report how much each work of art "moved" them. Intriguingly, the mind-wandering network lit up only when participants viewed artwork they *liked*—especially art that made them feel "touched from within."

The researchers concluded that "certain artworks, albeit unfamiliar, may be so well-matched to an individual's unique makeup that they obtain access to the neural substrates concerned with the self—access which other external stimuli normally do not get." More simply, *your experience of aesthetic pleasure is bound to your sense of personal identity.* The scientists added, "We propose that certain artworks can 'resonate' with an individual's sense of self in a manner that has well-defined physiological correlates and consequences, namely to regions of the default network." The sweet spots on your listener profile are a perfect example of this resonance between creative art and your innermost self.

The Nobel laureate Eric Kandel wrote that by linking the experience of one's personal self to the experience of art, the beholder actively *participates* in the artwork. When you view a painting, you evaluate whether the ideas and feelings it evokes within you match your self-concept. If you experience positive feelings while beholding a painting, then your mind-wandering network fires up, activating the circuits that support your sense of self, delivering a deeply rewarding experience: *This is the art of me.*

The listener's response is similar to the beholder's response, though they differ in one very important way. Our auditory circuitry has more varied and more direct connections to our emotion circuitry than does our visual circuitry. This is due to our brain's capacity for language, inarguably the most important mental tool in the *Homo sapiens* survival kit. Our need for better language capability drove the rapid development of our brain's melody network, lyrics network, rhythm network, and timbre network. As a result, we can detect the most subtle emotional shadings in a person's speech, automatically analyzing verbal intonations, rhythms, and word choices for the slightest hint of the speaker's true sentiments and intentions. Music listening exploits the same emotion-sensitive language circuitry.

Because of this, music activates our mind-wandering network— and our personal self—more easily and fully than any other art form.

4

So then. What happens in your brain, exactly, when you feel attracted to a record?

Robin Wilkins, a researcher at Wake Forest University in North Carolina, led a team that investigated people's brain activity as they listened to different genres of music, both familiar and unfamiliar. Participants relaxed in an fMRI scanner and listened on headphones to records in classical, country, rap, rock, and Chinese opera styles, plus the subject's self-chosen favorite song. The listeners rated each piece of music as Like, Dislike, or Favorite. Next, Wilkins examined participants' brain activity while they listened to the music. She observed interesting dynamics in a brain structure called the precuneus—a structure involved with self-awareness, self-consciousness, self-imagery, and creativity. The precuneus is *not* part of the mind-wandering network, though they are connected to each other.

The precuneus was activated in every listener's brain as they listened to every genre of music. When participants listened to music they rated as Like or Favorite, the activity between the precuneus and the mind-wandering network increased significantly. In contrast, when participants heard music they rated as Dislike, the precuneus stopped communicating with the mind-wandering network and was "connected primarily only to itself." This is astonishing. It suggests that our brain actively "rejects" disliked records, so when we hear music we don't enjoy, our brain automatically takes action to prevent those styles from getting integrated into our self-image.

Wilkins and her team made another surprising discovery. They detected significant communication between the auditory networks and the hippocampus (a brain structure involved in memory formation) whenever subjects listened to music they rated as Like or Dislike. That might have led the researchers to conclude that our hippocampus gets activated whenever we listen to *any* music, but Wilkins noticed that the communication between the auditory networks and the hippocampus was *reduced* when subjects listened to their "Favorite" music.

Wilkins and her team suggest that when we hear a favorite song, our memory circuits kick into *retrieval* mode rather than *encoding* mode and we automatically "play back" memories of people, places, and events that we associate with the song. Wilkins's intriguing hypothesis is supported by the research Ogi and I did on visualizations during music enjoyment: we found that the most common form of mental visualization that listeners experienced while listening to their favorite records was autobiographical memories.

These neuroscience findings suggest an answer to the big question that I posed at the beginning of this book: What is it about *you* that makes you feel the thrill of resonance when you hear one record but the chill of apathy when you hear another? More simply, what makes a person fall in love with a record?

The mind-wandering network is intricate and complex, touching many parts of the brain, including perceiving networks, thinking networks, feeling networks, and social networks. The mind-wandering network develops and evolves as these other networks learn through experience. Thus, your own mind-wandering network is intimately bound up with the unpredictable and singular arc of your life. Your mind-wandering network is the crucible of your one-of-a-kind listener profile.

The graphic on the opposite page adds the mind-wandering network to the "Musical Mind" graphic that we first encountered in the Timbre chapter. The new graphic illustrates how our music-processing circuitry is under the sway of our ever-drifting minds.

In large part, we seek out music for the rewards of letting our minds wander, rewards linked to our deepest conception of our personal self. Sometimes we need to access our most-buried emotions, while other times we need to feel our inner dancer, warrior, or athlete. Sometimes we need words to express our tangled thoughts; other times we'd like to visualize an impossible romance. We turn to our favorite records to take us where we want to go—where we *need* to go.

We don't choose our preferred musical "street" any more than we choose our height or sexual orientation. Our mind wanders where it will. All we can do is be open to learning what this wandering can teach us about ourselves. We should never deny our true nature, whether our musical love is for Billie Eilish, Billy Currington, Billy Idol, or Billie Holiday. In human relationships and musical relationships, your opportunities for fulfillment start with being honest with yourself about what you are truly attracted to.

Which brings us to my final and perhaps most important point about love at first sight. We are all prone to falling head over heels for someone, just because of the way they smiled, or talked, or moved across the room. When you feel that abrupt shock of attraction, your beloved seems perfect. But the truth of the matter is that

The Musical Mind

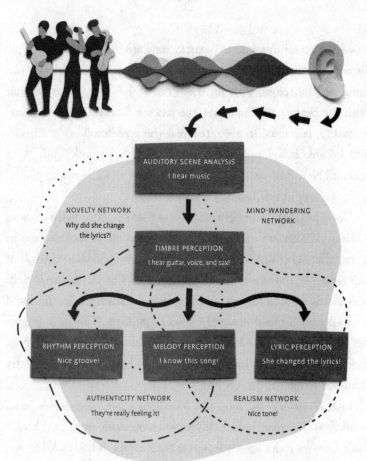

your beloved will always have flaws, perhaps major ones. They drive recklessly; they never admit mistakes; they say "catchup" instead of "ketchup"; they can be sneaky; they snort when they laugh. Your friends may express abject puzzlement that you could overlook such blemishes. But it just goes to show that human attraction always boils down to this truism:

Your beloved will never be perfect—but they can be perfect for you.

5

I owe my career in music to a particularly attentive awareness of my listener profile. The summer of my ninth birthday, an older cousin came to visit, along with his 45s. Prior to his visit, I was familiar with the Beatles, the Supremes, the Jackson 5, and most of what was played on pop radio in 1965, the year the skateboard and miniskirt first became fads. Cousin Mike dropped James Brown's "Papa's Got a Brand New Bag" on the turntable, and the music began to play.

It was love at first listen.

Upon hearing those tight horns and that staccato rhythm guitar, I felt a kid's version of *This is the s#&t!* The music spoke to me in a way that other records never had. The bass conducted the roll of our hips. The hi-hat told our feet what to do. The Godfather of Soul was so sure of what he was saying, so authoritative, leading his band and all who listened down the road to hipness. The funk groove felt like it vibrated the hidden core of my musical- and self-identity. It felt *resonant.* Music had never seemed so *right* before, so perfectly matched to me.

In that astonishing moment, I realized three things. First, soul music was the street I lived on. Second, pop music was not. Third, I hadn't *chosen* to set up residence on Soul Street. It had *called* to me, saying, "This is your home."

I wanted to experience that thrilling feeling of recognition and self-awareness again, so I searched for the same musical resonance in every record I heard. Whenever I scored some allowance or birthday money, it went toward new 45s, including Booker T. & the MGs, the Turtles, Buffalo Springfield, the Yardbirds, Marvin Gaye, and Creedence Clearwater Revival. Style was never the object of my pursuit. I was after a feeling of truth. From the start, I sought out music that transported me into a psychic landscape where my feelings and ideas could bloom.

I had no way of knowing at the time that this kind of attentiveness to my listener profile was laying the groundwork for becoming a record maker. It began a musical journey that seemed, then and now, like a vital and necessary relationship. I felt a *commitment* to music. I wanted to serve it, promote it, enjoy it, respect it, and be a good partner in every way I could. Love for music compelled me to study electronics to repair the equipment that recorded it, allowed me to work for days without sleep for my favorite artist, urged me to acquire the skills and confidence to become a record producer, led me into musical science and education—and drove me to write this book. I have never stopped embracing a love for music. For its own part, music has helped me to become more understanding of others, to make a living, and to delight with my students in its ever-changing form. Music shaped almost every aspect of my identity.

The musical street you live on when you're young won't necessarily change as you get older, but you might begin to visit other neighborhoods from time to time. Although the type of records you crave may vary, the records you enjoy when you're older usually have something in common with the records you enjoyed when you were younger, reflecting deep preferences that are part of your core.

Though I spent many years working and playing in the musical neighborhood of alternative rock, I always came home to soul music. Of course, when I was working in the industry, I was younger than I am now and was drawn to music that expressed my generation's interests and ideals. When I reached my forties, I strolled over to a new street as jazz became more and more compelling to me. Now that I'm in my sixties, I've been spending more time enjoying the blues. And every once in a while, I fall in love with a record that doesn't fit neatly into *any* of my preferred genres.

One day, my former student Ben Gebert (now of the band Haerts) played his production project in class. I've heard many hundreds of student records, but I had never experienced love at first listen until I heard Ben's recording of Russell Lacy's "Angelina."

The record starts with the hum of Russell's warm, rich vocals. Immediately I recognize the male vocal timbre I like best. Then the electronic piano and drums come in. The drummer taps out a light and even rhythm on brushes, not sticks, and the kick drum holds a soft and steady heartbeat while the snare dances in syncopation above it, almost like a hip-hop beat. I recognize my sweet spot on Rhythm Street. The Wurlitzer piano reminds me of the soul records I love so much, and the pianist sticks to that style, never showing off, adding grace notes and harmony around the chord progression. The sound is dry and present, almost uncomfortably so. The record sets a scene that is, to my ears, startlingly realistic, immediately allowing me the reward I get from visualizing the musicians. Within seconds, I feel this record might be special. I lean forward in anticipation.

And then, the thunderbolt of musical love! A lyric—authentic, original, inspired, poetic. Russell's opening lines:

> *Angelina, I will catch your breath*
> *Save it on my tongue*
> *Even though I am the lowest wretch*
> *For you I would atone.*

Russell continues singing to Angelina for a few verses but stops before finishing his thought. His musicians stop, too—except the drummer, who carries on alone. The feel is not like a typical drum break—it's more like the pregnant pause that happens at the scene of an accident. It is suspenseful and suggests unspoken truths.

> *Anywhere's a good place to die*
> *Any way is a good way to go.*

The close-mic'ing technique, along with the soft padding of the performances, gives a claustrophobic feel to the record. As Russell sings, I feel I'm in a small, tight room where something awful or unlawful has just happened, but I can't tell what's going on. I can see a room and a candle and a drink, but it's not clear where Angelina is or what has happened.

Angelina, we are bound
Songbirds to sound.

In my imagination, I hear the music echo in the hills of Virginia where Russell grew up, hills that have nurtured American songwriters for centuries. I can hear the timeless Americanization of immigrants in the sensibilities of two of my students who recently arrived in the United States, Ben and his wife Nini Fabi, as she blends her voice in the chorus with Russell's.

"Angelina" pairs the classic songwriter form I love with the imaginative lyric writing I crave. It is a realistic-sounding record that lets me visualize the performers and picture the story. It has a gentle, hummable melody that contrasts with its dark story, like a gorilla's hand in a lady's glove, yet they rise and fall together. The rhythm gets into the kind of pocket that my body most enjoys synchronizing to, and stays there.

I recognize the street I live on. I recognize *home*.

6

The story of music unfolds through the experiences of billions of listeners of every background and profile, each pursuing their own private musical love every time they play a record, attend a concert, write a rap, or program a drum beat. Out of these countless independent pursuits emerge large-scale trends of collective desire. Our individual quest for listener rewards makes music a dynamic ecosystem that is always in flux as each new generation of listeners joins the dance.

That's why I'd like to close this book the same way it opened:

with a record pull. Here are the voices of women and men from many backgrounds sharing their own memories of love at first listen. Each person describes, in their own words, one beloved record that accompanied them throughout their life. Each of these records generated profound feelings of appreciation, awe, and love in the listener who selected it—though it's unlikely these records will stir the same feelings in you. You have your own loves! Hopefully, as in any good record pull, hearing others explain why a particular record means so much to them will help you refine what you appreciate about the records you play over and over again.

T.J. is a middle-aged Californian, trim and agile enough to ride a skateboard and bold enough to still climb trees. He bought a house once owned by his grandparents, and he has a knack for capturing creative images with the camera on his phone. He says of his love-at-first-listen record:

"Her attitude comes shining through; she's rising up through the ensemble. She's saying to a lecherous man that she's not taking it anymore. She's in full sexuality and is not afraid of it. There is a weird torque, like seeing both the mosquito and the blood it seeks. The rumba box is primordial, it's volcanic, like the pulse of all humanity. When the drums come in we don't even need them, but the drums sound masculine while she is all feminine. It's both threatening and inviting, but it conjures up a feeling of who I am."

The record is "Funkier Than a Mosquito's Tweeter" by Nina Simone, 1974.

A.N. is just out of his twenties, and if there is someone more devoted to records, I've never had the pleasure to meet them. He suffered a traumatic head injury when he was in middle school but recovered fully, although as a young teen he had to learn to walk and feed himself all over again. He says of his love:

"As a kid I loved hardcore because it's kids saying whatever they want as aggressive and as hard as they want. But this song has these extreme dynamics and an atypical form, like classical; there's a sweetness in the writing and extreme dexterity in his voice. There is *movement* in this record; it feels real and emotional. It has a tangible urgency; you can hear that they all played it together. It sounds as if they had to record it right then or it never would have happened. I get a feeling of empathy for the performance but also a solitary feeling. There is so much *internal adventure* in there."

The record is "Mojo Pin" by Jeff Buckley, 1994.

J.B. is a jazz guitarist and young producer with an extraordinarily sensitive ear for music. In talking about music with him, you get the sense that as a listener he has fallen so deeply into it that he will never, could never, crawl free of it. Music is, for him, the only way of being in the world. He says of his love:

"A lot of times we listen for comfort, but it's also important to listen to music that makes you squirm. When I listen to this I feel seasick, I get chills, but for some bizarre reason, sometimes I chase that feeling. This song is about ending a relationship with someone. You know how hurtful it is—to yourself and to someone you care about—to have to break the news. The lyrics [of this song] work both ways. Either [the singer] is singing to you or you're the one singing it. When I first heard it, I was with the person I wanted to break up with. I don't know if she felt it, but the air was *thick*. I felt as if I was singing it through [the singer]. When I hear it now, the hairs on my arms stand up and I go back to that feeling—the air gets thick. All psychedelic music should have a sense of danger. We are trained to push away darkness, but I've been shaped by darkness as much as by beauty."

The record is "Eventually" by Tame Impala, 2015.

A.M. owns an art gallery in Los Angeles with his fiancée. His love of visual art extends to the artists who make it. After a decade in the business, his gratitude for the opportunity to bring deserving work to the public eye has only grown. In his free time, he plays the violin. He says of his love:

"If you close your eyes and listen for the full forty minutes, there is no way that you cannot cry. [He] wrote thirty variations—each with the same bass line—capped by two arias. The variations take you to places that rip your heart out, that send you soaring to heaven, and that are pedestrian. It is complicated to the point that I've been listening to it for fifteen years and I keep discovering new things. It's a window into a mind that is showing you how deep humans can be. Gould plays this version in a way of *knowing*—and he knows that you know—how great this is."

The record is Glenn Gould's 1981 recording of the Goldberg Variations, BWV 988, by J. S. Bach.

E.G. has a restless, adventurous appetite for travel, food, movies, and music. As a young man he left his native Venezuela to study music in the United States.

"I first heard this when I was fifteen years old. It destroyed me and built me up again, and today I am able to tap from this pure source of energy. This thing fell out of the sky and crushed everything. Up until that point, metal hadn't made a connection with African, funky rhythms. It was all Anglo, Germanic aggression. No one had made heavy stuff so damn groovy. When this kicks off, my body can't help but move. The guitars have a feeling like dragging your fingers down an inflated balloon—that sound that talks to your gut. This guitarist had glissandos; there was a slinky glissando thing that had a serpent-like movement. This was different from the Lego-block metal and was kind of nausea-inducing. It had a dissonance that was not very commonplace, counterbalanced with immaculate, superhuman, impossible precision. I still love it with my forty-four-year-old brain, but it brings me back to my fifteen-year-old self. It's more than a memory—I *become* fifteen again."

The record is "Mouth for War" by Pantera, 1992.

T.B. is an artist who grew up as an only child. He rebelled against the strictures of high school, and so he would escape to the library, where he spent his days scrutinizing books on painting and sculpting technique. He says of his love:

"Images played out in front of my eyes. This was the first record I *saw*. It's so profoundly personal, yet so odd. The lyrics and the sound of the voice, the backgrounds, the breaks—it's pop and yet it's not pop. It's extremely well constructed and yet it has a looseness. It felt like [the artist] talking to himself about something that mattered to him. For me it was an awakening in production. It opened up for me what music could *look* like. I was into surrealism at the time, and this record was like musical surrealism—familiar imagery in unfamiliar configurations."

The record is "I Am the Walrus" by the Beatles, 1967.

C.S. is a classically trained pianist born and raised in Japan. She is a heart-stopping performer, whether she is onstage as a solo artist or leading her avant-garde band. She says of her love:

"This record showed me the power of the studio. Every time I listen to it, I hear something different. Other albums I've aged out of, but this is a song that I keep coming back to. It is like a compass for the work I do. It has a broodiness, a darkness, and I'm drawn to dark things. It sounds like it is speaking almost from beyond the grave, like Emily Dickinson looking down on her own body. Dark music is an outlet, celebrating that we're still here. It is like a beacon; I keep falling in love all over again."

The record is "Religion" by Sheena Ringo, 2003.

W.M. is a rare bird—or, in her words, a "wild pony"—whose mannerisms and expressions make me think of music shaped into a human form. Her love affair with music started before she could talk and has never left her. She says:

"I was thirteen years old when I first heard this song. It completely destroyed my heart and changed who I became. My parents had divorced, and I was experiencing grief, loss, trouble, and anxiety. I didn't act out, but I had so much internal strife going on and I felt so much unfiltered impulsivity. When I heard this record it felt like I became an adult—it forced me to grow up—and I got a glimpse of what I could be when I got older. Her chords sounded like questions—suspended with no sense of the root. You were left to feel for yourself where it was coming from. Bob Dylan's words could make you feel like you were traveling, but this record gave me the sense of traveling through its chords."

The record is "Don Juan's Reckless Daughter" by Joni Mitchell, 1977.

A.R. works in commercial construction, and his life has had major ups and downs. He is extraordinarily tall and strong but so empathetic in nature that animals and children are magnetically drawn to him. He says of his musical love:

"The melody sounds like resurrection. It starts off describing a bad situation, but you can survive it. It has a great beginning, middle, and end. The feeling is solemn, but there is happiness and hope. There is an undercurrent of 'no anger.' I craved the even-keelness of it. I was always shy, introverted, and I hated confrontation. It was so easy to fall in love with that song. If I knew that a nuclear bomb was coming, I'd put on that record."

The record is "Brokedown Palace" (live in New Jersey) by the Grateful Dead, 1988.

THIS DIVERSITY OF MUSICAL romances both fuels and explains the origins of music itself. Our unique cravings serve as a bottomless source of inspiration.

When you put this book down, give your favorite record another listen. A close listen. No phones. No distractions. Tune in to those deep pathways that resonate inside you. When you start to focus on those special dimensions of music that make you swoon, without any preconceived idea of what constitutes good or cool or acceptable, you'll learn something real about yourself—and, perhaps, discover a new way to connect with others.

CODA

My coauthor and I kicked off our collaboration with a record pull. It didn't take long for us to realize that our tastes in music were completely different. Back and forth we went, playing songs one after another. It was tempting to want to impress each other, but that wouldn't have adhered to the spirit of a record pull. Record pulls are all about honesty. Ogi played a wide variety of records, but none matched my sweet spots. And though he was deeply curious about my musical choices, they didn't make his heart flutter, either. In most things musical, it turned out that we sought almost entirely opposite rewards.

During quiet moments in the studio, I enjoyed asking record makers to name a guilty pleasure—a record you would be embarrassed to admit you liked. Such confessions can be deeply revealing. The records we treasure covertly reflect facets of our musical self that we'd just as soon not have others know about. When it was my turn, I would admit to liking REO Speedwagon's "One Lonely Night," a sentimental ballad by a big-haired rock band that was particularly popular in the 1970s and '80s.

At the close of our record pull, Ogi and I knew enough about each other's apparent listener profiles that it was time to look a little deeper. What's your guilty pleasure? I shared "One Lonely Night." Next, Ogi sheepishly put on his own: "Cool Water," sung by Tim Blake Nelson.

At last! A connection!

Though Ogi and I live on different musical streets, during the waning moments of our record pull we finally discovered an unexpectedly joyous intersection: cowboy music. The music of Gene Autry, the Sons of the Pioneers, Montana Slim, and Marty Robbins. Even though our private fantasies carry us away to very different places when we listen to music, it turned out there was one kind of music that enabled us to share a similar pleasure.

Cowboy music, once called western music, feels different from its musical cousin, country music. For me, there is a solitary romance in cowboy music that places it in a niche all its own. Cowboy songs evoke a sense of loneliness wrapped within an optimistic and intrepid spirit. They feel like the songs a dog would sing if he could, and no other animal expresses loneliness quite the way a dog can. This record's simplicity seems to tap into the kind of pure love I feel when I am with animals. Listening to cowboy music, I smell the dust and feel the warmth rising from a horse's neck as we make our way to the next adventure.

For Ogi, cowboy music triggers visualizations of distant lands with vast, big-sky panoramas stretching from horizon to horizon, as a single traveler jauntily makes his way across the lonely chaparral. Although he's riding alone and is tiny against the backdrop of monumental Nature, the traveler's spirits remain undaunted and cheery, if perhaps tinted with a bit of wistful melancholy. Cowboy music always makes Ogi think of going on a journey into the unknown, perhaps even a journey into that ultimate unknown that awaits us all, but refusing to submit to sorrow, regret, or fear, and instead savoring memories of simple human pleasures—including the act of singing itself.

Because music can feel so intimate and personal, it is easy to forget that music is, above all, a form of sharing. As I remarked in the overture, music requires both a performer and a listener to exist at all. Composers, songwriters, and musicians connect with the music they heard in their youth, honing their listener profiles as they become increasingly sophisticated musical beings. As they begin to share their work, new listeners use it to develop their own listener profiles. Musical communiqués find their way into us, changing how we think, talk, move, dress, and interact with others.

It is fortunate for the art of music that we each respond to music in our own unique way. And it is fortunate for human beings that, on occasion, a record provides two people with the same jolt of delight, creating the opportunity for a connection that can run deeper than words.

ACKNOWLEDGMENTS

HILLARY CLINTON WAS ONCE ASKED ON CAMERA ABOUT the longevity of her partnership with her husband, Bill. She simply explained that they "started a conversation in the spring of 1971, and more than thirty years later, we're still talking." Susan can relate to that. She started a conversation about music with Tommy Jordan in the spring of 1992, and they are still talking. This book would not exist without him. Any insight or flash of brilliance glinting through its pages are ideas that germinated with Tommy.

Susan would also like to acknowledge her mentors and musical collaborators for their talents and what they taught her: Tony Berg, Jeff Black, Tim Bruckner, David Byrne, Lisa Coleman, Jim Creeggan, Robben Ford, Kevin Hearn, Jesse Johnson, Greg Kurstin, Nil Lara, George Massenburg, Wendy Melvoin, Craig Northey, Steven Page, Michael Penn, Sandy Roberton, Ed Robertson, Mark Rubel, Tyler Stewart, Al B. Sure, Tricky, Greg Wells, and Andrew Yeomanson. Special thanks to Todd Herreman for his assistance with song rights.

Every duo needs a tiebreaker. Each impasse between Susan and Ogi was instantly resolved by opening a sentence with the words "Tom thinks . . ." We couldn't have asked for a better editor than the peerless Tom Mayer, and no better publisher than Norton. We are thankful for art director Steve Attardo and the brilliant, vibrant, cover art by Mike Perry. That extra care and love helps this book sing. An exuberant holler to Bonnie Thompson, our copy editor, who "Sherlock Holmesed" every word and record in our book. We're grateful for the talents of Nneoma Amadi-obi, Elisabeth Kerr, Steve Colca, Anna Oler, Steven Pace, and Will Scarlett. We'd also like to give a major shout-out to our team at Levine Greenberg Rostan Literary Agency, including our excellent and much-appreciated agent Jim Levine, Courtney Paganelli, Michael Nardullo, and Melissa Rowland.

A royal mention is necessary for our brilliant illustrator, Hanna Piotrowska, who brought our ideas to visual life.

Susan and Ogi are grateful to the readers who reviewed our manuscripts and suggested many improvements: Nikolina Kulidžan, Rajeshwari Dutt, C. Crandall Hicks, John Davis, Katherine Rosenhammer, Steve Armin, Lori Dalton, Gail Schwartz, Robin Flinchum, Sarah Roman, Sai Gaddam, and Tofool Alghanem.

Susan wishes to thank the academic advisers and researchers whose commitment to understanding how music works makes them fountainheads in music cognition and auditory neuroscience. Their brilliance outshines the sun. They are: Albert Bregman, Peter Cariani, Nancy Etcoff, Tecumseh Fitch, Erica Knowles, Daniel Levitin, Psyche Loui, Stephen McAdams, Caroline Palmer, Aniruddh Patel, and Eve-Marie Quintin. She is also thankful to students, faculty, and staff at Berklee College of Music for the countless extracurricular conversations we've had about music and how best to make and serve it: Prince Charles Alexander, Carl Beatty, Jeremy Bernstein, Chad Blinman, Celia Bolgatz, Stephen Croes, Matthew Ellard, Enrique Gonzalez Müller, Jarred Hahn, Adam Moskowitz, Andrew Nault, Alex Prieto, Andrew Sarlo, Josh Sebek, Hank Shocklee, Sean Slade, Ebonie Smith, Courtney Swain, Barbara Thomas, John Whynot, and Steven Xia.

Finally, these acknowledgments would not be complete without admitting the heavy debt we owe to our loved ones. For Susan: the Rogers, Jordan, and Bruckner families, John Sacchetti, and "Cousin Mike" Van Meter. For Ogi: Tofool Alghanem and Zain Ogas.

NOTES

OVERTURE

1   **orange-and-crimson dragons:** Jimmy Page's black silk suit can be viewed on the Metropolitan Museum of Art's website: https://www.metmuseum.org/art/collection/search/754663.

10   **"listener profile" that is yours alone:** Cognitive psychologists use the term "cognitive profile" as a way of acknowledging that each human brain is one of a kind, shaped by the unfolding of trillions of microscopic biochemical events and years of experiences unique to you.

Chapter 1   AUTHENTICITY

20   **as one town hall regular described it:** Quoted in Susan Orlean's 1999 *New Yorker* article, "Meet the Shaggs."

21   **"ever recorded in the history of the universe":** Reprinted in John DeAngelis's liner notes for the Rounder Records compilation album *The Shaggs*, 1988.

23   **musical rules and theories:** In an interview, the designer Cecil Balmond declared, "That's an interesting 18th or 19th century concept by Friedrich Schiller. The naïve versus the *sentimental* is what is important. Those who try very hard for a particular thing, force themselves into it, and forcing it to work, is [*sic*] sentimental. In the naïve something else breaks through. Primitive art we call naïve, which doesn't mean that it's simplistic. True genius, like a Bach, or a Shakespeare, is naïve. Though the works are the ultimate in construction, they're naïve, because they come straight through to you, and enter into you. You take a Shakespeare play, and it's there [pointing at his chest]; it speaks to you directly. Whereas if you take a play by Marlowe, or someone else who is not such a great talent, what you recognize is that the author is working to make it work; you are conscious of layers of *trying* buried in the work; the work stays here [pointing to the head]. It's a kind of extreme argument, but it's interesting."

24   **"self-indulgent and cluttered with effects":** Vince Aletti reviewing Stevie Wonder's album *Where I'm Coming From* in *Rolling Stone* magazine, August 5, 1971. Wonder must have worked out the kinks soon after because he followed up this album with an unprecedented and practically impossible feat—he

released five great albums in a row: *Music of My Mind; Talking Book; Innervisions; Fulfillingness' First Finale*; and *Songs in the Key of Life*.

26 **legendary record producer Tony Berg:** Tony Berg makes records at Zeitgeist—his Los Angeles studio. He is a music purist in the sense that all of his creative projects aim to advance the state of the art, with less concern for pop chart success. In his decades-long career, Tony has influenced and mentored dozens of the finest and most successful producers, engineers, studio musicians, and artists to come out of Los Angeles. I was privileged to engineer several albums for him when I returned home after leaving Minnesota. Watching Tony taught me how to produce records.

26 **their own musical island:** Quoted in John DeAngelis, liner notes. *The Shaggs* (Rounder Records, 1988).

Chapter 2 REALISM

48 **figureless splashes of color:** The Nobel Prize–winning brain scientist Eric Kandel describes the strange way our brain perceives Pollock's paintings in his brilliant book *Reductionism in Art and Brain Science*: "His works do not have any points of emphasis or identifiable parts. They lack a central motif and encourage our peripheral vision. As a result, our eyes are constantly on the move: our gaze cannot settle or focus on the canvas. This is why we perceive action paintings as vital and dynamic." It's no coincidence that abstract works of art often resemble Rorschach ink blots—both are designed to elicit subjective responses that reveal the inner life of the viewer. The discussion of realism and abstraction in painting, including the impact of photography, is taken from ideas presented in Kandel's book.

Chapter 3 NOVELTY

67 **sales (including streaming):** The *Billboard* chart reflects US music consumption of all music in all formats, including sales, radio play, and streaming. Ryan Seacrest's American Top 40 is, by its proprietors' own admission, mostly based on adult contemporary records. *Billboard*'s greater inclusiveness is why it is the industry standard.

67 **follow along and learn:** The communications and information expert Christopher Burns notes on the *Point of Inquiry* podcast that humans evolved to be excellent learners. The problem is, we're really bad at *unlearning*.

68 **Hot 100 chart in 2019:** Famous musicians occasionally record music that is intentionally simplified in order to appeal to kids, such as Barenaked Ladies' *Snacktime!*, Peter Himmelman's *My Green Kite*, Ozomatli's *Ozokids*, David Grisman and Jerry Garcia's *Not for Kids Only*, and the wonderful *Family Time* by Ziggy Marley.

68 **presented in a familiar style:** Classic music styles can and do grow, though there is usually resistance from listeners who don't enjoy having their familiar musical forms stretched too far out of shape. Soundgarden and Pearl Jam emerged from the grunge music scene and found acceptance by fans of classic rock because they added originality in amounts that expanded the basic rock

blueprint in a way that was both timely and recognizable. A generation after Pearl Jam, the young Michigan rock band Greta Van Fleet has had a much tougher time finding widespread acceptance.

Greta Van Fleet has generated enviable media attention, and the group appears to possess all the ingredients of the next great rock band, including truly stunning vocals by lead singer Josh Kiszka that echo Robert Plant's. Nevertheless, rock musicians must obey a mandate for fearlessness. Rock music bred mosh pits and stage diving. When the Who's Pete Townshend destroyed his guitar at the end of a set, and when Nirvana's Kurt Cobain took a flying leap into Dave Grohl's drum kit, these performers demonstrated the fearlessness that defines rock music. Unfortunately, on one fateful night, Greta Van Fleet couldn't match their audacity. I watched with millions of others as the band performed on *Saturday Night Live* in early 2019. At the end of their first song, the guitar player did something odd. He made a move as though he was about to leap into the drum kit—but at the last moment he hesitated and pulled back. I felt a stab of empathetic pain. I knew how such an abortive move would look to rock fans. Sure enough, the next morning the music bloggers were not very kind.

72   **mocked for their unconventional tastes:** A man sitting across from me at a Boston dinner party accused me of "faking it" when I confessed that I wasn't a big fan of the Beatles and said that I preferred Sly & the Family Stone. The man said that my preference for Sly over John and Paul was merely a tactic for getting attention. That sad and puzzling incident oiled the mental gears that ultimately led to this book.

75   **extremely talented and highly trained:** You can hear Greg Kurstin's award-winning writing and production talents on a long list of popular hits, including Adele's "Hello" and "Easy on Me," Kelly Clarkson's "Stronger (What Doesn't Kill You)," and "Girl," by Maren Morris.

75   **such as circuit bending:** Circuit bending is the technique of partially dismantling a cheap electronic device (like a low-fi synthesizer or a child's toy) and rerouting or short-circuiting some of its components to create unusual and random sounds. It is a prominent feature of noise music and is sometimes performed on the live stage. To hear an example, try "Jacob's Ladder" by Arcane Device.

76   **launching the next trend:** You can watch a movement happen in under three minutes by viewing the YouTube video "First Follower: Leadership Lessons from Dancing Guy," which shows footage from an outdoor music festival (find a link at ThisIsWhatItSoundsLike.com). It is narrated by former Berklee student and founder of CD Baby, Derek Sivers. When the video begins, a shirtless dancing guy is the only concertgoer uninhibited enough to dance to the music. The rest of the listeners prefer to remain seated. But when a "first follower" is brave enough to cast his lot with Dancing Guy and join in, they are quickly followed by others. This is how a movement happens in musical trends, too, but on a longer time scale.

82   **adventure-seeking brains elsewhere:** This fact helped me understand the surprising popularity of "cover bands" in Boston—far greater than what I've observed in other major cities. For its size, Boston has an exceptionally high number of colleges and universities. Many overworked students enjoying a night out in a Boston bar cannot spare the bandwidth to focus on and enjoy original, unfamiliar tunes.

Chapter 4 MELODY

93 **Ahmanson Theatre in Los Angeles:** Readers can hear this performance in the documentary *Sinatra: All or Nothing at All* (2015). It's available through several popular streaming services.

98 **the same chord progression:** A chord is a set of three or more notes played simultaneously. A sequence of chords is known as a chord progression—a sort of "tonal skeleton" for a song. Taylor Swift's "The 1" starts with a repeating two-chord progression on piano that is joined by another chord progression on guitar before Swift's vocal enters to sing the melody.

107 **before we were even born:** Musical elements can be interpreted one way by listeners from a given generation or culture, and another way by listeners from a different generation or culture. Samuel Barber's *Adagio for Strings* (you may know it as the theme from the movie *Platoon*) has been called the saddest melody ever written. Although Barber was frequently melancholy, nothing suggests that he thought the composition was sad. Barber composed the *Adagio* in the summer of 1936 during a particularly happy time in his life. After completing the work, he wrote to a friend, "I have just finished the slow movement of my quartet today—it is a knockout!" Music history professor Luke Howard writes in *American Music* that Barber had no intention for the piece to be used as a funeral hymn. It became America's "semi-official music for mourning" after it was broadcast following the deaths of Franklin Delano Roosevelt and John Fitzgerald Kennedy. Today it is a cultural model of what "sadness" sounds like in music.

108 **and other rock music:** In *Sound Targets: American Soldiers and Music in the Iraq War*, Jonathan Pieslak states that some claimed that the tactic was not intended to unnerve Noriega as much as it was to boost the morale of American troops.

110 **called the Shepard tone:** Musical pitches have two properties: pitch chroma (note names) and pitch height. The twelve chromatic pitches (called semitones) on a piano are duplicated seven times across seven full octaves. Even in non-human animals, frequencies in a 2:1 ratio (an octave) are perceived by the brain as equivalent. This inspired the psychoacoustician Roger Shepard to plot pitch chroma and pitch height as an ascending spiral, circling around in a helix such that every chroma was directly above the identical chroma an octave lower. This circular property lets a set of computer-generated pitches create an auditory illusion.

If we present to a listener the twelve adjacent semitones in series, one after the other, amplifying one tone at a time, it gives the overall impression of a steadily rising pitch. After hearing twelve tones, we are an octave higher than where we began and, due to "octave equivalence," perceptually back at the beginning. Just like artist M. C. Escher's famous visual illusion of a never-ending staircase, Shepard tones seem to rise and rise without ever reaching the top.

Chapter 5 LYRICS

117 **story in a song's lyrics:** Studies with poetry (Belfi, Vessel, and Starr, 2017) show that vividness of imagery is the most significant factor in a poem's appeal, and

this could very well apply to music lyrics, too, given that imagining the story is a common music-listening activity.

118 **rise of social media:** This was the only time I witnessed a mail drop while working with Prince. His management soon arranged to have his fan mail delivered elsewhere, perhaps because these overflowing bags were the very emblem of the word "futility." There was simply no way to keep up with the influx.

119 **content (the *information*):** The more a voice is altered to sound like an instrument, such as when it's heavily filtered through a vocoder or Auto-Tune, the less our brain processes the signal in its "singing area" and the more we hear the voice as an instrument (Lévêque and Schön, 2015).

Chapter 6   RHYTHM

148 **measure of tone deafness:** If you're curious about the MBEA, you can take the test online for free: http://musicianbrain.com/mbea/. It's kind of fun!

152 **sped up or slowed down:** Researchers Ani Patel and research assistants visited Snowball in Indiana to test his capacity for beat induction. They manipulated the tempo of musical excerpts across a wide range and found that Snowball spontaneously adjusted the tempo of his rhythmic movements to stay in sync (Patel, Iversen, Bregman, and Schulz, 2009).

156 **the extraction of a tactus:** The interested reader may enjoy this review paper by W. Tecumseh Fitch; his work is both scholarly and wonderful: W. Tecumseh Fitch, "Four Principles of Bio-musicology," *Philosophical Transactions of the Royal Society B: Biological Sciences* 370, no. 1664 (2015): 20140091. https://doi.org/10.1098/rstb.2014.0091.

157 **to be an "Elaine":** Elaine Benes was a character on the television comedy *Seinfeld* (1989–98). In one famous episode, she is revealed to be a risibly bad dancer.

162 **tap of the snare drum:** In "Poinciana," the drummer disengages the "snares" underneath the snare drum to make it sound more like a tom.

Chapter 7   TIMBRE

171 **except in certain cases of alliteration:** One type of alliteration refers to the end of a sound or word being identical to the start of the next word in the sequence. Lyric writers can put an alliteration into a lyric and add a clever spin to the text. For example, Geggy Tah's "Sacred Cow" features the lyric "What song reminds you of when / Life was home on the dangerous / Which side of the tracks are you on?" Singer Tommy Jordan alliterates the end of "dangerous" and the beginning of "Which," making it sound like "switch side."

171 **"continuous beat" is nonsensical:** The Risset accelerando effect is an auditory illusion that sounds like a continually accelerating tempo. The effect is caused by an automatic process of perceptual reorganization. It is named after Jean-Claude Risset, a pioneer of computer music. As the drum beat seems to get faster and faster, eventually it reaches a rate that our experienced brains believe is impossible or, at the very least, highly unlikely. So the brain automatically reorganizes the incoming audio stream to regroup the percussive hits into larger "chunks," effectively slowing down the tempo.

176 **powerful emotional effect on us:** Compared with our perception of melody, lyrics, and rhythm, timbre perception is especially complex, not only due to timbre's infinite variety but, as we saw with the comparison of old and new violins, its ties to our prior experiences with a sound. Timbre reveals the identity of a sound source, and many sources of sound are attention-grabbing, such as a shrieking baby, a flirty whisper, an approaching siren, or the ding of a text message. We learn to categorize sounds as rewarding or punishing, relevant or irrelevant, promising or deflating, depending on what we've experienced with them in the past.

180 **excites your eardrum:** Nearly every sound we hear arrives through the ear canals, but there are also bone-conducted sounds and internally generated audible sounds.

182 **"a chirping sparrow":** Alluri et al.: "Timbral features activated mainly perceptual and resting-state or default mode areas of the cerebrum and cognitive areas of the cerebellum. In contrast, for tonal and rhythmic features, we observed for the first time during listening to a naturalistic stimulus, activations in subcortical emotion-related areas along with activations in cognitive and somatomotor cerebrocortical areas."

183 **pleasure—or displeasure:** Independent processing networks can account for why the right song heard at the wrong time can feel aversive or unwelcome. You may love every single musical element of your favorite record, but if you hear it while you and your beloved are arguing over money, it won't sound the same. Your aesthetic circuitry is alerting your reward network: Inappropriate!

184 **caused them to react:** We learn to recognize the timbre of different instruments as we hear them in childhood. The term for this acquired knowledge is "timbre template," and it applies to vocals, too (Handel and Erickson, 2004).

Chapter 8 FORM AND FUNCTION

199 **"analytic listening":** In the music-cognition community, the word "musician" describes a person with at least five years of formal musical training that began in childhood. Musical practice in childhood, when our brains are most pliable, leads to physical changes: the entire auditory pathway gets thicker and more robust, growing additional "branches" on the auditory nerves to better distinguish among similar sounds.

Trained musicians are better able to listen *analytically* to chords: their auditory system hears and recognizes each note that makes up a chord. This is similar to the ability of "supertasters" who can sample a spoonful of soup and name every ingredient. Nonmusicians generally lack the ability to isolate the notes in a chord. They listen *synthetically*: they hear a chord as a single sonic object.

You can listen to an Acoustical Society of America audio demonstration called "ASA 25—Analytic vs. Synthetic Pitch," provided by the Correlogram Museum, to learn if you hear analytically or synthetically. (Find a link at ThisIs WhatItSoundsLike.com.) You will hear a pair of tones presented against a background of noise. Do you hear the pitch go up or down? If you hear it go down, you are listening analytically. If you hear the pitch go up, you are listening synthetically. Why the difference?

The first tone of the pair has two frequencies: 1,000 and 800 Hertz (Hz).

The second tone also has two frequencies: 1,000 and 750 Hz. Analytic listeners hear the 800 Hz frequency component drop down to 750 Hz, like a falling pitch. Synthetic listeners, in contrast, hear the pitch *rise* from 200 Hz to 250 Hz. The pitch of the first tone sounds like 200 Hz to synthetic listeners, corresponding to the spacing between the simultaneous frequencies of 1,000 and 800 Hz. The second tone sounds like a pitch of 250 Hz, corresponding to the spacing between 1,000 and 750 Hz.

209 **"Little Red Corvette":** "Crossover single" is the term for a song that earns enough sales and radio play to appear not only on a specialized chart but also on the *Billboard* Hot 100. Many a great single has first appeared on the rock, country, or R&B charts but never achieved the airplay to cross over to the Hot 100.

216 **as they did on every album:** The "naked song" on *Stunt* was called "Contrary," but it didn't make the final cut. We had recorded most of the album and were just getting started on "Contrary" when drummer Tyler said, "Hey, guys, we haven't done the naked song yet." No sooner was the "na" out of his mouth than all five band members' clothes came off. (They may have left their shoes on.) We did the usual number of takes—five or six—this way and then I beckoned them into the control room. I assumed they'd put their clothes back on, but when I looked up from the console, five naked men were aligned in a row behind me. They listened to the playback and commented on it as usual, then marched to the front office to ask the studio manager for change for a dollar. I could have cried for joy. They were treating me like they treated each other. I was one of them.

BIBLIOGRAPHY

LINER NOTES

Bannister, Scott. "A Vigilance Explanation of Musical Chills? Effects of Loudness and Brightness Manipulations." *Music & Science* 3 (2020). https://doi.org/2059204320915654.

Belfi, A. M., and P. Loui. "Musical Anhedonia and Rewards of Music Listening: Current Advances and a Proposed Model." *Annals of the New York Academy of Sciences* 1464, no. 1 (2020): 99–114.

Castro, São Luís, and César F. Lima. "Age and Musical Expertise Influence Emotion Recognition in Music." *Music Perception: An Interdisciplinary Journal* 32, no. 2 (2014): 125–42.

Kirschner, Sebastian, and Michael Tomasello. "Joint Music Making Promotes Prosocial Behavior in 4-Year-Old Children." *Evolution and Human Behavior* 31, no. 5 (2010): 354–64.

Levitin, Daniel J., and Susan E. Rogers. "Absolute Pitch: Perception, Coding, and Controversies." *Trends in Cognitive Sciences* 9, no. 1 (2005): 26–33. https://doi.org/10.1016/j.tics.2004.11.007.

Mas-Herrero, Ernest, et al. "Dissociation Between Musical and Monetary Reward Responses in Specific Musical Anhedonia." *Current Biology* 24, no. 6 (2014): 699–704.

Mas-Herrero, Ernest, et al. "Individual Differences in Music Reward Experiences." *Music Perception: An Interdisciplinary Journal* 31, no. 2 (2012): 118–38.

Panksepp, Jaak. "The Emotional Sources of 'Chills' Induced by Music." *Music Perception* 13, no. 2 (1995): 171–207. https://doi.org/10.2307/40285693.

Patel, Aniruddh D., et al. "Speech Intonation Perception Deficits in Musical Tone Deafness (Congenital Amusia)." *Music Perception* 25, no. 4 (2008): 357–68. https://doi.org/10.1525/mp.2008.25.4.357.

Peretz, Isabelle, and Dominique T. Vuvan. "Prevalence of Congenital Amusia." *European Journal of Human Genetics* 25, no. 5 (2017): 625–30. https://doi.org/10.1038/ejhg.2017.15.

Sanes, Dan H., and Sarah M. N. Woolley. "A Behavioral Framework to Guide Research on Central Auditory Development and Plasticity." *Neuron* 72, no. 6 (2011): 912–29.

Serafine, Mary Louise. *Music as Cognition: The Development of Thought in Sound.* New York: Columbia University Press, 1988.

Zamm, Anna, et al. "Pathways to Seeing Music: Enhanced Structural Connectivity in Colored-Music Synesthesia." *Neuroimage* 74 (2013): 359–66. https://doi.org/10.1016/j.neuroimage.2013.02.024.

Chapter 1   AUTHENTICITY

Anderson, Thomas. "In the Studio with the Shaggs." *Blurt.* Published online 2016. Accessed February 2020. https://blurtonline.com/feature/in-the-studio-with-the-shaggs/.

Balmond, Cecil, and Eric Ellingsen. "Survival Patterns." In *Models*, vol. 11, edited by Emily Abruzzo, Eric Ellingsen, and Jonathan D. Solomon. New York: 306090 Books, 2007. (Page 27.)

Christoff, Kalina, et al. "Mind-Wandering as Spontaneous Thought: A Dynamic Framework." *Nature Reviews Neuroscience* 17, no. 11 (2016): 718–31.

Chusid, Irwin. "The Shaggs: Groove Is in the Heart." In *Songs in the Key of Z: The Curious Universe of Outsider Music*, 4–11. Chicago: A Capella Books, 2000.

DeAngelis, John. Liner notes. *The Shaggs*. Rounder CD11547. 1988.

Dickinson, Emily. Letter to Thomas Higginson. L459a. 1876. Reproduced in *The Complete Poems of Emily Dickinson*. New York: Little, Brown, 1976.

Fishman, Howard. "The Shaggs Reunion Concert Was Unsettling, Beautiful, Eerie, and Will Probably Never Happen Again." *New Yorker*, August 30, 2017. https://www.newyorker.com/culture/culture-desk/the-shaggs-reunion-concert-was-unsettling-beautiful-eerie-and-will-probably-never-happen-again.

Grant, B. Rosemary, and Peter R. Grant. "Songs of Darwin's Finches Diverge When a New Species Enters the Community." *Proceedings of the National Academy of Sciences* 107, no. 47 (2010): 20156–63.

Orlean, Susan. "Meet the Shaggs." *New Yorker*, September 22, 1999. https://www.newyorker.com/magazine/1999/09/27/meet-the-shaggs.

Ronson, Jon. *Jon Ronson On.* Series 6, episode 3, "The Fine Line Between Good and Bad: The Shaggs." Broadcast June 6, 2002; produced by Lucy Greenwell for White Pebble Media and Renegade Pictures, BBC 4. https://www.bbc.co.uk/programmes/b010y002.

Solomon, Jonathan D., Emily Abruzzo, and Eric Ellingsen, eds. *Models*. Vol. 11 of *306090*. Princeton, NJ: Princeton Architectural Press, 2008.

Chapter 2   REALISM

Aviv, Vered. "What Does the Brain Tell Us About Abstract Art?" *Frontiers in Human Neuroscience* 8 (2014): 85. https://doi.org/10.3389/fnhum.2014.00085.

Christoff, Kalina, et al. "Mind-Wandering as Spontaneous Thought: A Dynamic Framework." *Nature Reviews Neuroscience* 17, no. 11 (2016): 718–31. https://doi.org/10.1038/nrn.2016.113.

Danto, Arthur C. *The Madonna of the Future: Essays in a Pluralistic Art World*. Berkeley: University of California Press, 2001.

Delaney, Darby. "How Martin Scorsese Perfected the Movie Soundtrack." *Film School Rejects*, July 17, 2018. https://filmschoolrejects.com/how-martin-scorsese-perfected-the-movie-soundtrack/.

Frost, Robert. "Dust of Snow." *New Hampshire*. New York: Henry Holt, 1923.

Galassi, Peter. *Before Photography: Painting and the Invention of Photography*. New York: Morgan Press, 1981. (Page 12.)

Gombrich, Ernst H. *The Essential Gombrich: Selected Writings on Art and Culture*. Edited by Richard Woodfield. London: Phaidon, 1996. (Page 108.)

Kandel, Eric R. *Reductionism in Art and Brain Science: Bridging the Two Cultures*. New York: Columbia University Press, 2016.

Kawabata, Hideaki, and Semir Zeki. "Neural Correlates of Beauty." *Journal of Neurophysiology* 91, no. 4 (2004): 1699–1705. https://doi.org/10.1152/jn.00696.2003.

McDonald, John. "James Turrell: A Retrospective." *Sydney Morning Herald*. February 15, 2015.

Mesquita, Batja, Lisa Feldman Barrett, and Eliot R. Smith, eds. *The Mind in Context*. Guilford, 2010.

Moyle, Franny. *Turner: The Extraordinary Life and Momentous Times of J.M.W. Turner*. New York: Penguin Press, 2016.

Rose, Todd. *The End of Average: How to Succeed in a World That Values Sameness*. New York: HarperOne, 2016.

Rose, Todd, and Ogi Ogas. *Dark Horse: Achieving Success Through the Pursuit of Fulfillment*. New York: HarperCollins, 2018.

Rose, L. Todd, Parisa Rouhani, and Kurt W. Fischer. "The Science of the Individual." *Mind, Brain, and Education* 7, no. 3 (2013): 152–58.

Salle, David. *How to See*. New York: W. W. Norton, 2016. (Page 23.)

Schwartz, Sanford. *Artists and Writers*. New York: Yarrow, 1990. (Page 203.)

Turrell, James. "Aten Reign at the Guggenheim and James Turrell's Skyspaces." Interview posted on YouTube, December 8, 2016. https://www.youtube.com/watch?v=_rW0N7B5KD4.

———. Interview posted on jamesturrell.com, 2021. https://jamesturrell.com/about/introduction/.

Chapter 3 NOVELTY

Carpentier, Sarah M., et al. "Complexity Matching: Brain Signals Mirror Environment Information Patterns During Music Listening and Reward." *Journal of Cognitive Neuroscience* 32, no. 4 (2020): 734–45. https://doi.org/10.1162/jocn_a_01508.

Chmiel, Anthony, and Emery Schubert. "Back to the Inverted-U for Music Preference: A Review of the Literature." *Psychology of Music* 45, no. 6 (2017): 886–909. https://doi.org/10.1177/0305735617697507.

Ferreri, Laura, et al. "Dopamine Modulates the Reward Experiences Elicited by Music." *Proceedings of the National Academy of Sciences* 116, no. 9 (2019): 3793–98. https://doi.org/10.1073/pnas.1811878116.

Marin, Manuela M., et al. "Berlyne Revisited: Evidence for the Multifaceted Nature of Hedonic Tone in the Appreciation of Paintings and Music." *Frontiers in Human Neuroscience* 10 (2016): 536. https://doi.org/10.3389/fnhum.2016.00536.

Medawar, Peter B. "The Threat and the Glory." In *The Threat and the Glory*. New York: HarperCollins, 1990.

Nicholson, Nigel, et al. "Personality and Domain-Specific Risk Taking." *Journal of Risk Research* 8, no. 2 (2005): 157–76. https://doi.org/10.1080/1366987032000123856.

Percino, Gamaliel, Peter Klimek, and Stefan Thurner. "Instrumentational Complexity of Music Genres and Why Simplicity Sells." *PLOS One* 9, no. 12 (2014): e115255. https://doi.org/10.1371/journal.pone.0115255.

Ridenhour, Carlton. *Chuck D Presents: This Day in Rap and Hip-Hop History.* New York: Hachette, 2017. (Page 7.)

Salimpoor, Valorie N., et al. "Interactions Between the Nucleus Accumbens and Auditory Cortices Predict Music Reward Value." *Science* 340, no. 6129 (2013): 216–19. https://doi.org/10.1126/science.1231059.

Salimpoor, Valorie N., et al. "Predictions and the Brain: How Musical Sounds Become Rewarding." *Trends in Cognitive Sciences* 19, no. 2 (2015): 86–91. https://doi.org/10.1016/j.tics.2014.12.001.

Sallavanti, Micalena I., Vanessa E. Szilagyi, and Edward J. Crawley. "The Role of Complexity in Music Uses." *Psychology of Music* 44, no. 4 (2016): 757–68. https://doi.org/10.1177/0305735615591843.

Sapolsky, Robert M. *Behave: The Biology of Humans at Our Best and Worst.* New York: Penguin, 2017. (Pages 161–68.)

Serrà, Joan, et al. "Measuring the Evolution of Contemporary Western Popular Music." *Scientific Reports* 2, no. 1 (2012): 1–6. https://doi.org/10.1038/srep00521.

Zuckerman, Marvin. "The Sensation Seeking Scale V (SSS-V): Still Reliable and Valid." *Personality and Individual Differences* 43, no. 5 (2007): 1303–05. https://doi.org/10.1016/j.paid.2007.03.021.

Zuckerman, Marvin, and D. Michael Kuhlman. "Personality and Risk-Taking: Common Bisocial Factors." *Journal of Personality* 68, no. 6 (2000): 999–1029. https://doi.org/10.1111/1467-6494.00124.

Chapter 4 MELODY

Bernstein, Leonard. *Leonard Bernstein's Young People's Concerts.* New York: Anchor, 1962. (Page 201.)

Brown, Steven. "Are Music and Language Homologues?" *Annals of the New York Academy of Sciences* 930, no. 1 (2001): 372–74. https://doi.org/10.1111/j.1749-6632.2001.tb05745.x.

Deutsch, Diana, Trevor Henthorn, and Rachael Lapidis. "Illusory Transformation from Speech to Song." *Journal of the Acoustical Society of America* 129, no. 4 (2011): 2245–52. doi.org/10.1121/1.3562174.

"The 500 Greatest Albums of All Time." *Rolling Stone* (October 2020): 41–89. Published online September 22, 2020. https://www.rollingstone.com/music/music-lists/best-albums-of-all-time-1062063/.

Grossberg, Stephen, et al. "ARTSTREAM: A Neural Network Model of Auditory Scene Analysis and Source Segregation." *Neural Networks* 17, no. 4 (2004): 511–36. https://doi.org/10.1016/j.neunet.2003.10.002.

Hardach, Sophie. "Do Babies Cry in Different Languages?" *New York Times*, November 14, 2019. Published online April 15, 2020. Accessed June 10, 2021. https://www.nytimes.com/2020/04/15/parenting/baby/wermke-prespeech-development-wurzburg.html.

Howard, Luke. "The Popular Reception of Samuel Barber's *Adagio for Strings.*" *American Music* (2007): 50–80.

Kaplan, James. *Frank: The Voice.* New York: Anchor, 2011. (Pages 105–07.)

Mampe, Birgit, et al. "Newborns' Cry Melody Is Shaped by Their Native Language." *Current Biology* 19, no. 23 (2009): 1994–97. https://doi.org/10.1016/j.cub.2009.09 .064.

McConnell, Patricia B. "Lessons from Animal Trainers: The Effect of Acoustic Structure on an Animal's Response." In *Perspectives in Ethology*, vol. 9, edited by P.P.G. Bateson and Peter H. Klopfer. New York: Plenum Press, 1991.

McDermott, Josh H. "Auditory Preferences and Aesthetics: Music, Voices, and Everyday Sounds." In *Neuroscience of Preference and Choice*, edited by Raymond Dolan and Tali Sharot, 227–56. Waltham, MA: Academic Press, 2012. https://doi.org/10 .1016/B978-0-12-381431-9.00020-6.

Nummenmaa, Lauri, Vesa Putkinen, and Mikko Sams. "Social Pleasures of Music." *Current Opinion in Behavioral Sciences* 39 (2021): 196–202. https://doi.org/10.1016/j .cobeha.2021.03.026.

Orenstein, Arbie. *Ravel: Man and Musician*. New York: Dover, 1991. (Page 98.)

Owings, Donald H., and Eugene S. Morton. *Animal Vocal Communication: A New Approach*. New York: Cambridge University Press, 1998.

Patel, Aniruddh D., John R. Iversen, and Jason C. Rosenberg. "Comparing the Rhythm and Melody of Speech and Music: The Case of British English and French." *Journal of the Acoustical Society of America* 119, no. 5 (2006): 3034–47. https://doi.org/10.1121/1.2179657.

Pieslak, Jonathan R. *Sound Targets: American Soldiers and Music in the Iraq War*. Bloomington: Indiana University Press, 2009. (Pages 82–86.)

Ross, Deborah, Jonathan Choi, and Dale Purves. "Musical Intervals in Speech." *Proceedings of the National Academy of Sciences* 104, no. 23 (2007): 9852–57. https:// doi.org/10.1073/pnas.0703140104.

Seyfarth, Robert M., and Dorothy L. Cheney. "Production, Usage, and Comprehension in Animal Vocalizations." *Brain and Language* 115, no. 1 (2010): 92–100. doi. org/10.1016/j.bandl.2009.10.003.

Snowdon, Charles T., and David Teie. "Affective Responses in Tamarins Elicited by Species-Specific Music." *Biology Letters* 6, no. 1 (2010): 30–32. https://doi.org/10 .1098/rsbl.2009.0593.

———. "Emotional Communication in Monkeys: Music to Their Ears?" In *Evolution of Emotional Communication: From Sounds in Nonhuman Mammals to Speech and Music in Man*, edited by Eckart Altenmüller, Sabine Schmidt, and Elke Zimmermann, 133–51. Oxford, UK: Oxford University Press, 2013.

Soley, Gaye, and Erin E. Hannon. "Infants Prefer the Musical Meter of Their Own Culture: A Cross-Cultural Comparison." *Developmental Psychology* 46, no. 1 (2010): 286. https://doi.org/10.1037/a0017555.

Wermke, Kathleen, and Werner Mende. "Musical Elements in Human Infants' Cries: In the Beginning Is the Melody." *Musicae Scientiae* (Supplement) 13, no. 2 (2009): 151–75. https://doi.org/10.1177/1029864909013002081.

Chapter 5 LYRICS

Amodio, David M., and Chris D. Frith. "Meeting of Minds: The Medial Frontal Cortex and Social Cognition." *Nature Reviews Neuroscience* 7, no. 4 (2006): 268–77. https://doi.org/10.1038/nrn1884.

Appel, Nadav. "'Ga, ga, ooh-la-la': The Childlike Use of Language in Pop-Rock Music." *Popular Music* 33, no. 1 (2014): 91–108. https://www.jstor.org/stable/24736973.

Axelrod, Jim. "Journey's 'Don't Stop Believin'' Turns 30." CBS News. Published online June 5, 2012. https://www.cbsnews.com/news/journeys-dont-stop-believin -turns-30/.

Belfi, Amy M., et al. "Rapid Timing of Musical Aesthetic Judgments." *Journal of Experimental Psychology: General* 147, no. 10 (2018): 1531. https://doi.org/10.1037/ xge0000474.

Belfi, Amy M., Edward Vessel, and G. Gabrielle Starr. "Individual Ratings of Vividness Predict Aesthetic Appeal in Poetry." *Psychology of Aesthetics Creativity and the Arts* 12, no. 3 (2017). https://doi.org/10.1037/aca0000153.

Bizley, Jennifer K., and Yale E. Cohen. "The What, Where and How of Auditory-Object Perception." *Nature Reviews Neuroscience* 14, no. 10 (2013): 693–707. https://doi.org/10.1038/nrn3565.

Bono. "60 Songs That Saved My Life." *Rolling Stone.* Published online May 15, 2020.

Borčak, Lea Wierød. "The Sound of Nonsense: On the Function of Nonsense Words in Pop Songs." *SoundEffects: An Interdisciplinary Journal of Sound and Sound Experience* 7, no. 1 (2017): 27–43. https://doi.org/10.7146/se.v7i1.97177.

Brinkley, Douglas. "Don McLean's 'American Pie.'" Christie's. February 2015. Accessed June 9, 2021. https://www.christies.com/lot/lot-don-mclean-b1945-the -complete-working-manuscript-5885030/?from=salesummary&intObjectID=5885 030&lid=1.

Christenson, Peter G., et al. "What Has America Been Singing About? Trends in Themes in the U.S. Top-40 Songs: 1960–2010." *Psychology of Music* 47, no. 2 (2019): 194–212. https://doi.org/10.1177/0305735617748205.

Fingerhut, Joerg, et al. "The Aesthetic Self: The Importance of Aesthetic Taste in Music and Art for Our Perceived Identity." *Frontiers in Psychology* 11 (2021): 4079. https://doi.org/10.3389/fpsyg.2020.577703.

Frith, Simon. "Music and Identity." In *Questions of Cultural Identity*, edited by Stuart Hall and Paul du Gay, 108–27. Thousand Oaks, CA: Sage, 1996.

Gill, A. A. "America the Marvelous." *Vanity Fair.* June 14, 2013. Published online July 2013. Accessed May 14, 2020. https://www.vanityfair.com/culture/2013/07/ america-with-love-aa-gill-excerpt.

Greenberg, David M., et al. "The Self-Congruity Effect of Music." *Journal of Personality and Social Psychology* 121, no. 1 (2020): 137–50. https://doi.org/10.1037/ pspp0000293.

Greene, Andy. "Steve Perry: 5 Songs That Inspired Me." *Rolling Stone.* Published online October 24, 2018.

Gunther Moor, Bregtje, et al. "Do You Like Me? Neural Correlates of Social Evaluation and Developmental Trajectories." *Social Neuroscience* 5, nos. 5–6 (2010): 461–82. https://doi.org/10.1080/17470910903526155.

Herd, Denise. "Changing Images of Violence in Rap Music Lyrics: 1979–1997." *Journal of Public Health Policy* 30, no. 4 (2009): 395–406. https://doi.org/10.1057/jphp .2009.36.

History by Day. "The Complete True Story Behind 'American Pie' by Don McLean." Published online (date unknown). Accessed June 9, 2021. https://www .historybyday.com/pop-culture/the-complete-true-story-behind-american-pie-by -don-mclean-3/39.html.

Janata, Petr. "The Neural Architecture of Music-Evoked Autobiographical Memories." *Cerebral Cortex* 19, no. 11 (2009): 2579–94. https://doi.org/10.1093/cercor/bhp008.

Lévêque, Yohana, and Daniele Schön. "Modulation of the Motor Cortex During Singing-Voice Perception." *Neuropsychologia* 70 (2015): 58–63.

Murphey, Tim. "The When, Where, and Who of Pop Lyrics: The Listener's Prerogative." *Popular Music* 8, no. 2 (1989): 185–93. https://www.jstor.org/stable/853468.

Nawrocki, Tom. "Rewind: The Biggest Instrumental Hits of the Past 50 Years." *Cuepoint*, April 10, 2015. Accessed November 10, 2021. https://medium.com/cuepoint/what-do-the-harlem-shake-star-wars-gary-glitter-hawaii-five-o-and-barry-white-have-in-common-542dc7c0c545.

Nummenmaa, Lauri, Vesa Putkinen, and Mikko Sams. "Social Pleasures of Music." *Current Opinion in Behavioral Sciences* 39 (2021): 196–202. https://doi.org/10.1016/j.cobeha.2021.03.026.

Peretz, Isabelle, and Max Coltheart. "Modularity of Music Processing." *Nature Neuroscience* 6, no. 7 (2003): 688–91. https://doi.org/10.1038/nn1083.

Recording Industry Association of America. "Top 365 Songs of the Twentieth Century." Published in March 2001 by the Recording Industry Association of America (RIAA) and the National Endowment for the Arts (NEA). http://www.theassociation.net/txt-music5.html.

Sapolsky, Robert M. *Behave: The Biology of Humans at Our Best and Worst.* New York: Penguin, 2017. (Page 165.)

Schlaug, Gottfried, et al. "From Singing to Speaking: Facilitating Recovery from Nonfluent Aphasia." *Future Neurology* 5, no. 5 (2010): 657–65. https://doi/10.2217/fnl.10.44.

Schwartz, John. "To Know Me, Know My iPod." *New York Times*, November 28, 2004. https://www.nytimes.com/2004/11/28/weekinreview/to-know-me-know-my-ipod.html.

Smirke, Richard. "U2 Producer Andy Barlow on 'Songs of Experience': 'The Album Changed Massively After Trump Got Elected.'" *Billboard*, June 12, 2017. https://www.billboard.com/music/rock/andy-barlow-interview-u2-producer-songs-experience-8061774/.

"Top 100 Instrumental Songs Since 1960." Tunecaster. Published online (date unknown). Accessed November 18, 2021. http://tunecaster.com/special/most-popular/instrumentals.html.

Ventzislavov, Rossen. "Singing Nonsense." *New Literary History* 45, no. 3 (2014): 507–22. https://doi/10.1353/nlh.2014.0024.

Zatorre, Robert J., Pascal Belin, and Virginia B. Penhune. "Structure and Function of Auditory Cortex: Music and Speech." *Trends in Cognitive Sciences* 6, no. 1 (2002): 37–46. https://doi.org/10.1016/S1364-6613(00)01816-7.

Zatorre, Robert J., Joyce L. Chen, and Virginia B. Penhune. "When the Brain Plays Music: Auditory–Motor Interactions in Music Perception and Production." *Nature Reviews Neuroscience* 8, no. 7 (2007): 547–58. https://doi.org/10.1038/nrn2152.

Chapter 6 RHYTHM

Cook, Peter, et al. "A California Sea Lion (*Zalophus californianus*) Can Keep the Beat: Motor Entrainment to Rhythmic Auditory Stimuli in a Non Vocal Mimic." *Journal of Comparative Psychology* 127, no. 4 (2013): 412–27.

Drake, Carolyn, Mari Riess Jones, and Clarisse Baruch. "The Development of Rhythmic Attending in Auditory Sequences: Attunement, Referent Period, Focal Attending." *Cognition* 77, no. 3 (2000): 251–88.

Dreifus, Claudia. "Exploring Music's Hold on the Mind." *New York Times*, May 31, 2010.

Fitch, W. Tecumseh. "The Biology and Evolution of Rhythm: Unravelling a Paradox." In *Language and Music as Cognitive Systems*, edited by Patrick Rebuschat et al. Oxford, UK: Oxford University Press, 2012.

———. "Dance, Music, Meter and Groove: A Forgotten Partnership." *Frontiers in Human Neuroscience* 10 (2016): 64.

———. "Four Principles of Bio-musicology." *Philosophical Transactions of the Royal Society B: Biological Sciences* 370, no. 1664 (2015): 20140091.

———. "Rhythmic Cognition in Humans and Animals: Distinguishing Meter and Pulse Perception." *Frontiers in Systems Neuroscience* 7 (2013): 68.

Good, Arla, and Frank A. Russo. "Singing Promotes Cooperation in a Diverse Group of Children." *Social Psychology* 47, no. 6 (2016): 340–44.

Guralnick, Peter. *Sam Phillips: The Man Who Invented Rock 'n' Roll.* New York: Back Bay, 2015. (Pages 15 and 255.)

Hattori, Yuko, Masaki Tomonaga, and Tetsuro Matsuzawa. "Spontaneous Synchronized Tapping to an Auditory Rhythm in a Chimpanzee." *Scientific Reports* 3, no. 1 (2013): 1–6.

Honing, Henkjan. "Without It No Music: Beat Induction as a Fundamental Musical Trait." *Annals of the New York Academy of Sciences* 1252, no. 1 (2012): 85–91.

Iversen, John R., Bruno Repp, and Aniruddh Patel. "Top-Down Control of Rhythm Perception Modulates Early Auditory Responses." *Annals of the New York Academy of Sciences* 1169, no. 1 (2009): 58–73.

Keehn, R. Joanne Jao, et al. "Spontaneity and Diversity of Movement to Music Are Not Uniquely Human." *Current Biology* 29, no. 13 (2019): R621–22.

Koelsch, Stefan, Peter Vuust, and Karl Friston. "Predictive Processes and the Peculiar Case of Music." *Trends in Cognitive Sciences* 23, no. 1 (2019): 63–77.

Large, Edward W., and Patricia M. Gray. "Spontaneous Tempo and Rhythmic Entrainment in a Bonobo (*Pan paniscus*)." *Journal of Comparative Psychology* 129, no. 4 (2015): 317.

Lindner, Axel, et al. "Human Posterior Parietal Cortex Plans Where to Reach and What to Avoid." *Journal of Neuroscience* 30, no. 35 (2010): 11715–25.

MacDougall, H. G., and S. T. Moore. Marching to the Beat of the Same Drummer: The Spontaneous Tempo of Human Locomotion. *Journal of Applied Physiology* 99, no. 3 (2005): 1164–73.

Martens, Peter A. "The Ambiguous Tactus: Tempo, Subdivision Benefit, and Three Listener Strategies." *Music Perception: An Interdisciplinary Journal* 28, no. 5 (2011): 433–48.

Mathias, Brian, et al. "Electrical Brain Responses to Beat Irregularities in Two Cases of Beat Deafness." *Frontiers in Neuroscience* 10 (2016): 40.

McAuley, J. Devin, et al. "The Time of Our Lives: Life Span Development of Timing and Event Tracking." *Journal of Experimental Psychology: General* 135, no. 3 (2006): 348.

Merchant, Hugo, and Apostolos P. Georgopoulos. "Neurophysiology of Perceptual and Motor Aspects of Interception." *Journal of Neurophysiology* 95, no. 1 (2006): 1–13.

Montagu, Jeremy. "How Music and Instruments Began: A Brief Overview of the Origin and Entire Development of Music, from Its Earliest Stages." *Frontiers in Sociology* 2 (2017): 8.

Patel, Aniruddh D. *Music, Language, and the Brain.* New York: Oxford University Press, 2010.

Patel, Aniruddh D., and John R. Iversen. "The Evolutionary Neuroscience of Musical Beat Perception: The Action Simulation for Auditory Prediction (ASAP) Hypothesis." *Frontiers in Systems Neuroscience* 8 (2014): 57.

Patel, Aniruddh D., John R. Iversen, Micah R. Bregman, and Irena Schulz. "Experimental Evidence for Synchronization to a Musical Beat in a Nonhuman Animal." *Current Biology* 19, no. 10 (2009): 827–30.

Pearce, Eiluned, Jacques Launay, and Robin I. M. Dunbar. "The Ice-Breaker Effect: Singing Mediates Fast Social Bonding." *Royal Society Open Science* 2, no. 10 (2015): 150221.

Phillips-Silver, Jessica, et al. "Born to Dance but Beat Deaf: A New Form of Congenital Amusia." *Neuropsychologia* 49, no. 5 (2011): 961–69.

Richter, Joachim, and Roya Ostovar. "'It Don't Mean a Thing If It Ain't Got That Swing': An Alternative Concept for Understanding the Evolution of Dance and Music in Human Beings." *Frontiers in Human Neuroscience* 10 (2016): 485.

Ross, Jessica M., John R. Iversen, and Ramesh Balasubramaniam. "The Role of Posterior Parietal Cortex in Beat-Based Timing Perception: A Continuous Theta Burst Stimulation Study." *Journal of Cognitive Neuroscience* 30, no. 5 (2018): 634–43.

Sisario, Ben. "Charlie Watts, the Unlikely Soul of the Rolling Stones." *New York Times*, August 24, 2021. https://www.nytimes.com/2021/08/24/arts/music/charlie-watts-rolling-stones.html.

Temperley, David. "Communicative Pressure and the Evolution of Musical Styles." *Music Perception* 21, no. 3 (2004): 313–37.

Witek, Maria A. G., et al. "Syncopation, Body-Movement and Pleasure in Groove Music." *PLOS One* 9, no. 4 (2014): e94446.

Yong, Ed. "Not a Human, but a Dancer." *Atlantic*, July 8, 2019. Accessed November 21, 2021. https://www.theatlantic.com/science/archive/2019/07/what-snowball-dancing-parrot-tells-us-about-dance/593428/.

Zentner, Marcel, and Tuomas Eerola. "Rhythmic engagement with music in infancy." *Proceedings of the National Academy of Sciences* 107, no. 13 (2010): 5768–73.

Chapter 7 TIMBRE

Abdurraqib, Hanif. "The TR-808 Drum Machine Changed the Sound of Pop Music Forever." *Smithsonian*, July 2020.

Abrams, Daniel A., et al. "Auditory Brainstem Timing Predicts Cerebral Asymmetry for Speech." *Journal of Neuroscience* 26, no. 43 (2006): 11131–37.

Alluri, Vinoo, Petri Toiviainen, Iiro P. Jääskeläinen, Enrico Glerean, Mikko Sams, and Elvira Brattico. "Large-Scale Brain Networks Emerge from Dynamic Processing of Musical Timbre, Key and Rhythm." *NeuroImage* 59 (2012): 3677–89.

Arnal, Luc H., et al. "Human Screams Occupy a Privileged Niche in the Communication Soundscape." *Current Biology* 25, no. 15 (2015): 2051–56.

Barratt, Emma L., and Nick J. Davis. "Autonomous Sensory Meridian Response (ASMR): A Flow-like Mental State." *PeerJ* 3 (2015): e851.

Bernstein, David. "The Moog Synthesizer Makes a Comeback." *New York Times*, September 29, 2004.

Bregman, Albert S. *Auditory Scene Analysis: The Perceptual Organization of Sound.* Cambridge, MA: MIT Press, 1990.

Collins, Sarah A. "Men's Voices and Women's Choices." *Animal Behaviour* 60, no. 6 (2000): 773–80.

Collins, Sarah A., and Caroline Missing. "Vocal and Visual Attractiveness Are Related in Women." *Animal Behaviour* 65, no. 5 (2003): 997–1004.

Erickson, Molly L., and Susan R. Perry. "Can Listeners Hear Who Is Singing? A Comparison of Three-Note and Six-Note Discrimination Tasks." *Journal of Voice* 17, no. 3 (2003): 353–69.

Feinberg, David R., et al. "The Role of Femininity and Averageness of Voice Pitch in Aesthetic Judgments of Women's Voices." *Perception* 37, no. 4 (2008): 615–23.

Fritz, Claudia, et al. "Player Preferences Among New and Old Violins." *Proceedings of the National Academy of Sciences* 109, no. 3 (2012): 760–63.

Fritz, Claudia, et al. "Soloist Evaluations of Six Old Italian and Six New Violins." *Proceedings of the National Academy of Sciences* 111, no. 20 (2014): 7224–29.

Gibson, Caitlin. "A Whisper, Then Tingles, Then 87 Million YouTube Views: Meet the Star of ASMR." *Washington Post*, December 15, 2014.

Grossberg, Stephen. "Adaptive Resonance Theory: How a Brain Learns to Consciously Attend, Learn, and Recognize a Changing World." *Neural Networks* 37 (2013): 1–47.

Handel, Stephen, and Molly L. Erickson. "Sound Source Identification: The Possible Role of Timbre Transformations." *Music Perception* 21, no. 4 (2004): 587–610.

Hughes, Susan M., Franco Dispenza, and Gordon G. Gallup Jr. "Ratings of Voice Attractiveness Predict Sexual Behavior and Body Configuration." *Evolution and Human Behavior* 25, no. 5 (2004): 295–304.

Kumar, Sukhbinder, et al. "The Brain Basis for Misophonia." *Current Biology* 27, no. 4 (2017): 527–33.

Mantione, Mariska, Martijn Figee, and Damiaan Denys. "A Case of Musical Preference for Johnny Cash Following Deep Brain Stimulation of the Nucleus Accumbens." *Frontiers in Behavioral Neuroscience* 8 (2014): 152.

McAdams, Stephen. "Recognition of Sound Sources and Events." *Thinking in Sound: The Cognitive Psychology of Human Audition* (1993): 146–98.

McDermott, Josh H. "Auditory Preferences and Aesthetics: Music, Voices, and Everyday Sounds." In *Neuroscience of Preference and Choice*, edited by Raymond Dolan and Tali Sharot, 227–56. Waltham, MA: Academic Press, 2012.

Nagyvary, Joseph, et al. "Wood Used by Stradivari and Guarneri." *Nature* 444, no. 7119 (2006): 565.

Peretz, Isabelle. "Towards a Neurobiology of Musical Emotions." In *Handbook of Music and Emotion: Theory, Research, Applications*, edited by Patrik N. Juslin and John A. Sloboda, 99–126. Oxford, UK: Oxford University Press, 2011.

Poerio, Giulia Lara, et al. "More Than a Feeling: Autonomous Sensory Meridian Response (ASMR) Is Characterized by Reliable Changes in Affect and Physiology." *PLOS One* 13, no. 6 (2018): e0196645.

Puts, David Andrew, Steven J. C. Gaulin, and Katherine Verdolini. "Dominance and the Evolution of Sexual Dimorphism in Human Voice Pitch." *Evolution and Human Behavior* 27, no. 4 (2006): 283–96.

Rickly, Geoff. Interview with Trent Reznor. *Alternative Press*. June 26, 2004.

Rouw, Romke, and Mercede Erfanian. "A Large-Scale Study of Misophonia." *Journal of Clinical Psychology* 74, no. 3 (2018): 453–79.

Salimpoor, Valorie N., et al. "Predictions and the Brain: How Musical Sounds Become Rewarding." *Trends in Cognitive Sciences* 19, no. 2 (2015): 86–91.

Sapolsky, Robert M. *Behave: The Biology of Humans at Our Best and Worst.* New York: Penguin, 2017. (Page 41.)

Scapelliti, Christopher. "The Guitar Gear Behind Derek & the Dominos' 'Layla.'" *Guitar Player.* July 22, 2020.

Stoel, Berend C., and Terry M. Borman. "A Comparison of Wood Density Between Classical Cremonese and Modern Violins." *PLOS One* 3, no. 7 (2008): e2554.

Tai, Hwan-Ching, et al. "Acoustic Evolution of Old Italian Violins from Amati to Stradivari." *Proceedings of the National Academy of Sciences* 115, no. 23 (2018): 5926–31.

Chapter 8 FORM AND FUNCTION

Birmingham Times. "Ten of the Greatest Jazz Groups, Bands, and Orchestras." June 29, 2016. http://www.birminghamtimes.com/2016/06/10-of-the-greatest-jazz-groups-bands-orchestras/.

Gallucci, Michael. "Grateful Dead Albums Ranked Worst to Best." Ultimate Classic Rock. June 24, 2015. https://ultimateclassicrock.com/grateful-dead-albums-ranked/.

Guralnick, Peter. *Sam Phillips: The Man Who Invented Rock 'n' Roll.* New York: Back Bay, 2015. (Page 166.)

Hall, Rick. *The Man from Muscle Shoals: My Journey from Shame to Fame.* Monterey: Heritage Builders, 2015. (Page 187.)

Holden, Stephen. "Pop: Prince, a Renegade." *New York Times*, March 28, 1981.

Joyce, Mike. "Robben Ford 'Supernatural' Blue Thumb." *Washington Post*, November 12, 1999.

Myers, Paul. *Barenaked Ladies: Public Stunts, Private Stories.* New York: Simon and Schuster, 2007.

Smoorenburg, Guido F. "Pitch Perception of Two-Frequency Stimuli." *Journal of the Acoustical Society of America* 48, no. 4B (1970): 924–42.

Chapter 9 FALLING IN LOVE

Belfi, Amy M., et al. "Rapid Timing of Musical Aesthetic Judgments." *Journal of Experimental Psychology: General* 147, no. 10 (2018): 1531.

Berridge, Kent C., Terry E. Robinson, and J. Wayne Aldridge. "Dissecting Components of Reward: 'Liking,' 'Wanting,' and Learning." *Current Opinion in Pharmacology* 9, no. 1 (2009): 65–73.

Brielmann, Aenne A., and Denis G. Pelli. "Beauty Requires Thought." *Current Biology* 27, no. 10 (2017): 1506–13.

Christoff, Kalina, et al. "Mind-Wandering as Spontaneous Thought: A Dynamic Framework." *Nature Reviews Neuroscience* 17, no. 11 (2016): 718–31.

Davis, Miles, and Quincy Troupe. *Miles.* New York: Simon and Schuster, 1990. (Page 333.)

James, William. *The Principles of Psychology.* New York: Henry Holt, 1890.

Kandel, Eric. *Reductionism in Art and Brain Science*. New York: Columbia University Press, 2016.

Miu, Andrei C., Simina Pițur, and Aurora Szentágotai-Tătar. "Aesthetic Emotions Across Arts: A Comparison Between Painting and Music." *Frontiers in Psychology* 6 (2016): 1951.

Salimpoor, Valorie N., et al. "Predictions and the Brain: How Musical Sounds Become Rewarding." *Trends in Cognitive Sciences* 19, no. 2 (2015): 86–91.

Vessel, Edward A., G. Gabrielle Starr, and Nava Rubin. "Art Reaches Within: Aesthetic Experience, the Self and the Default Mode Network." *Frontiers in Neuroscience* 7 (2013): 258.

Wilkins, Robin W., et al. "Network Science and the Effects of Music Preference on Functional Brain Connectivity: From Beethoven to Eminem." *Scientific Reports* 4, no. 1 (2014): 1–8.

Zsok, Florian, et al. "What Kind of Love Is Love at First Sight? An Empirical Investigation." *Personal Relationships* 24, no. 4 (2017): 869–85.